ANATOMY & PHYSIOLOGY
A complete introduction

David Le Vay

Revised and updated by
Jennifer Stafford-Brown

First published in Great Britain in 1996 by Hodder Education. An Hachette UK company.

First published in the USA in 1974 by The McGraw-Hill Companies, Inc.

This edition published in Great Britain in 2015 by John Murray Learning

This edition published in the USA in 2015 by Quercus

Copyright © David Le Vay 1974, 1985, 2001, 2003, 2015

British Library Cataloguing in Publication Data: a catalogue record for this title is available from the British Library.

Library of Congress Catalog Card Number: on file.

Paperback ISBN 978 1 47360866 5

eBook ISBN 978 1 47360867 2

5

Cover image © iStockPhoto.com

Typeset by Cenveo® Publisher Services.

Printed and bound in Great Britain by CPI Group (UK) Ltd., Croydon, CR0 4YY.

John Murray Learning policy is to use papers that are natural, renewable and recyclable products and made from wood grown in sustainable forests. The logging and manufacturing processes are expected to conform to the environmental regulations of the country of origin.

Carmelite House

50 Victoria Embankment

London EC4Y 0DZ

www.hodder.co.uk

ANATOMY & PHYSIOLOGY
A complete introduction

Contents

Acknowledgements

I wish to thank the following for permission to reproduce illustrations and other material: Longmans Green & Co. as regards *Gray's Anatomy* and J. & A. Churchill in respect of Starling's *Principles of Human Physiology*. I have also drawn widely on *A Companion to Medical Studies*, ed. R. Passmore and J.S. Robson, Blackwell Scientific Publications. The sources of these and other figures are acknowledged in the text. I am also grateful to the Oxford University Press for permission to reproduce Figure 2.4 from *Here Be Dragons* by D. Koerner and S. LeVay.

Part 1:

Anatomy

1

General considerations

In this chapter you will learn about:

▶ *what anatomy means*

▶ *key terms used in anatomy*

▶ *different methods of acquiring anatomical knowledge*

▶ *the history of anatomy.*

Anatomy

The word *anatomy* means the cutting up of the body to examine its parts. Knowledge gained in this way is essentially *regional* in that one gains a familiarity with each body part, such as the arm or leg. However, every part contains the same *kinds* of blood vessels, nerves, bones and so on – so that, as well as regional anatomy, there is also a *systemic* aspect in which the body is considered to be made up of several coordinated systems such as the nervous system. Body systems are known as *macroscopic* or *gross* anatomy, and contrast with *microscopic* anatomy or histology, which is the study of the structure of cells and tissues.

We should also consider *developmental* anatomy, which is known as *ontogeny*, and consists of *embryology*, which is the growth of the individual within the womb, *postnatal* development from infancy to maturity, and then, towards the end of life, certain changes as one grows older called *senescence*.

Methods of acquiring anatomical knowledge

▶ Dissection

The oldest method of acquiring anatomical knowledge is by *dissection*, which involves cutting body tissue and examining it. This can take place on both the living body and the recently dead through *surgical explorations* and *autopsies* respectively.

Nugget

In the early nineteenth century it was difficult to obtain subjects for dissection; only the bodies of hanged criminals were available – in fact, their dissection was compulsory by law. However, as this source proved inadequate, the snatching of freshly buried corpses by the 'resurrection men' became prevalent in London, Glasgow and Edinburgh, a lucrative occupation. Increasing public anger was aroused, especially when murders were committed to obtain bodies for sale, as in the infamous Burke and Hare episode in Edinburgh. The disclosures at their trial in 1828 were ultimately responsible for the Anatomy Act of 1831, regularizing the conduct of schools of anatomy and authorizing the use of donated bodies for dissection.

▶ Optical microscopy

This method uses electron microscopes to study microscopic anatomy of the cells and tissues which links up with the study of *molecular biology*.

▶ Surface anatomy inspection

These methods deal with the relation of superficial landmarks to the deeper structures in the body. Inspection reveals the bulge of muscles when they contract, the pulsation of arteries and the course of veins, and the position of bony prominences. *Palpation*, *manipulation* and *percussion* are methods used in surface anatomy to reveal respectively the consistence of the deep structures, the movements of joints, and the boundaries of hollow air-containing or solid organs, while listening with a stethoscope – *auscultation* – locates organs such as the heart, lungs and bowel.

▶ **Endoscopy**

This is the introduction of an instrument carrying an optical system, which may be connected to a videocamera, to visualize the inside of cavities such as the abdomen or hollow organs like the stomach or bladder, or the interior of joints (*arthroscopy*).

▶ **Ultrasound**

This method produces images of deep-lying structures without irradiation and is routinely used for studying the foetus in pregnancy; it shows up most internal organs but without much detail.

▶ **Radiographic anatomy**

X-rays show the skeletal system and also demonstrate hollow organs when these are filled with air, which is abnormally translucent, or with substances opaque to the rays.

▶ **Computed tomography (CT)**

This method uses X-rays and produces 'slices' at selected depths and planes of various body parts and areas.

▶ **Magnetic resonance imaging (MRI)**

MRI is a magnetic excitation of tissue hydrogen molecules without irradiation and also allows sections of any desired thickness, at any depth and in any plane of areas of the body, but remains a lengthy procedure.

Variation in human anatomy

Human beings are essentially similar, but there is a continuous variation from the standard pattern in inessential details. We vary outwardly in height and colouring, and internally there may be minor differences in the arrangement of nerves and blood vessels, bile ducts and bronchi. But just as the general catalogue – 'item, two lips, indifferent red; item, two eyes, with lids to them; item, one neck, one chin', and so forth – is always correct, so the femoral artery or the biceps muscle is always where we expect to find it. There are occasional grosser errors which represent a failure or misdirection of normal development, such as absence of part or whole of a limb or cleft palate.

The history of anatomy

Some prehistoric anatomical instinct is evident from the uncanny accuracy of carvings and cave drawings and indicates an acquaintance with underlying structures. The Egyptians had a body of knowledge derived from embalming, various surgical procedures and ritual divination using the entrails of animals. However, it is to the Greeks that we owe the idea of structure as a matter for deliberate investigation. The superb surface anatomy of Greek sculpture speaks for itself and various operations, such as trephining for head injuries, were common. **Aristotle** and **Hippocrates** are the great names of this period. Hippocrates worked from a practical surgical viewpoint in connection with wounds, fractures and dislocations. Aristotle, who founded comparative anatomy, did not practise human dissection but made many accurate observations of adult and embryo animals. After his death, the school at Alexandria established a discipline which included the public dissection of human bodies. There was a medical school in Rome and systematic descriptions of bones and organs began. In the second century AD, **Galen** was the greatest physician and anatomist of antiquity; his work was the basis of European anatomy for a thousand years, surviving the Dark Ages to reappear in translations by Arab scholars at the Renaissance.

The Middle Ages saw the rise of the great Italian universities and medical schools – Salerno, Bologna and Padua – where students were formally instructed in anatomy by professors. **Leonardo da Vinci** is remarkable in this period as a brilliant artist and anatomist who established principles of anatomic illustration which were developed by the invention of printing. In the sixteenth century, **Vesalius** replaced the traditional reliance on Galen by direct personal observation and dissection; his great work, *On the Fabric of the Human Body*, rejuvenated anatomy. In England, there seems to have been no dissection until the end of the fifteenth century, but some demonstrations on the corpses of criminals by the barber-surgeons were authorized under Henry VIII. During the latter part of the eighteenth century, the brothers **William** and **John Hunter** gave an immense impetus, not only to anatomy, but to its applications in surgery. John's specimens became the nucleus of the great Hunterian Collection at the Royal College of Surgeons. By the close of the century anatomy was recognized as a basic introductory subject for medical students.

Test yourself

1 Which of the following is an example of systemic anatomy?
 a The heart
 b An arm
 c The nervous system
 d A kidney

2 The term used for developmental anatomy is:
 a Regional
 b Postnatal
 c Ontogeny
 d Biology

3 What term is used in anatomy for a person as they are getting older?
 a Senescence
 b Ontogeny
 c Palpation
 d Morphology

4 Which of the following methods of acquiring anatomical knowledge involves cutting body tissues?
 a Endoscopy
 b Ultrasound
 c Dissection
 d Optical microscopy

5 Which of the following methods of acquiring anatomical knowledge involves using an optical instrument to look inside joints?
 a Endoscopy
 b Athroscopy
 c Optical microscopy
 d Radiographic anatomy

6 Which of the following methods of acquiring anatomical knowledge involves using magnetic excitation of tissue hydrogen molecules without irradiation?
 a Dissection
 b Palpation
 c Computed tomography
 d Magnetic resonance imaging

7 Which of the following methods of acquiring anatomical knowledge involves using inspection of superficial landmarks of the body?
 a Palpation
 b Endsocopy
 c Ultrasound
 d Optical microscopy

8 Which of the following people did *not* practise human dissection?
    a Aristotle
    b Hippocrates
    c Galen
    d Vesalius

9 Which of the following people was a brilliant artist?
    a Galen
    b Leonardo da Vinci
    c John Hunter
    d Aristotle

10 What did the Greeks use trephining for?
    a Head injuries
    b Arm injuries
    c Chest injuries
    d Abdominal injuries

2

Cells, DNA and tissues

In this chapter you will learn about:

▶ *the different parts of a cell*

▶ *the function of DNA*

▶ *the different types of tissues.*

Cells

A *cell* can be seen using an electron microscope after it has been stained with dyes. Only the largest cell, the fertilized ovum, is just visible to the naked eye.

All the body cells have sprung from the division of the female germ cell (ovum) after fertilization by the male germ cell (spermatozoön). Different cells have widely differing life-spans. Some are rapidly replaced, others are very long-lived; but all (except the germ cells, which, given the chance, are immortal) are programmed for ageing (senescence) and death. Some, such as white blood cells and certain cells of the immune system, can move around – either carried passively in the bloodstream or migrating actively like amoebae through the tissues – but most remain where they are formed.

In the adult there are:

▶ specialized cells that cannot divide and are irreplaceable, e.g. nerve and muscle cells

▶ cells that divide slowly in health but are stimulated to rapid growth in case of need, e.g. connective tissue in repair after injury

▶ cells that reproduce rapidly to replace those with a short life-span, e.g. the precursors of red blood cells and the outer layer of the skin.

Though the general tendency of a cell is to be spherical, many cells have bizarre shapes. Some nerve cells, though minute, send their fibres almost the length of the body. Muscles consist of elements often as long as the muscle itself. The red blood cell is modified into a flattened biconcave disc to concentrate the pigment haemoglobin at its periphery.

The essential parts of any cell are:

▶ the *bounding membrane*

▶ the *cytoplasm*

▶ the central *nucleus*.

▶ The bounding membrane

This is made up of lipids and proteins and is typically *semi-permeable,* allowing the passage of water and certain simple ions. It is not a mere envelope but an actively selective agent. Where cells are closely packed together, connecting channels allow communication of materials. At the membrane there is a resting negative electrical potential of −20 to −200 mV, i.e. an electrochemical gradient based on the different concentrations of sodium and potassium ions within and outside the cell.

The *surface* of each cell has characteristics of the tissue, the species and the individual. It contains proteins which reject foreign material from other organisms, i.e. it has an *immunological* function. It also possesses specialized receptors which 'lock on' to chemicals such as hormones (Chapter 21) and drugs which recognize other cells by direct contact or by electrical and chemical signalling.

▶ The cytoplasm

The *cytoplasm* is all the material between the nucleus and the bounding membrane. It is a mass of colloidal proteins, carbohydrates and smaller molecules, and contains most of the

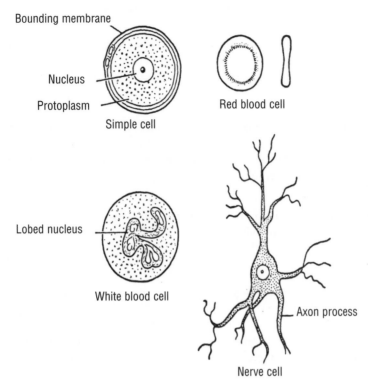

Figure 2.1 Various types of cells, showing that they can have very different shapes

ribonucleic acid (RNA) of the cell. It also contains a number of small discrete structures, or *organelles*, of various kinds.

▶ The *Golgi apparatus*, near the nucleus, is a stack of discoid structures which receives and reroutes proteins manufactured in the cytoplasm.

▶ The *mitochondria* are thread-like bodies. There are thousands of these in a cell and they contain enzymes for the oxidation of carbohydrates. They are the main sites for energy conversion and are prominent in very active tissues such as heart muscle and kidney tubules. They contain deoxyribonucleic acid (DNA) and RNA and replicate by simple fission.

▶ The *lysosomes* contain enzymes concerned with digesting absorbed material and scavenging.

▶ The *ribosomes* are dense particles of RNA-protein complexes and as such are sites of production of protein molecules.

▶ The *centrioles* are a pair of barrel-shaped bodies at right angles to each other near the nucleus, which have an important function in cell division.

▶ The *microtubules* radiate out from the nucleus and maintain cell shape with other organelles strung out along them.

▶ The *endoplasmic reticulum* is a network of fine tubes and membranes studded with ribosomes providing a large area for synthesis of proteins and cell membrane constituents.

▶ The nucleus

The *nucleus* is sharply defined and contains *chromosomes* (see below) with RNA and long molecules of DNA. Certain very large cells are multinucleate, e.g. bone osteoclasts and skeletal muscle fibres. Every body cell has a nucleus except the mature red blood cell, which has a very limited life because without nuclei cells cannot form proteins, which are needed in order to survive.

Cells stick together and this facilitates their arrangement as tissues. Cancer cells, however, lose this adhesion and then invade the body.

Some cells have the power of *phagocytosis*, i.e. they can surround and ingest foreign matter – bacteria and dead cells; this is notable in the polymorphs of the blood and the cells of connective tissue.

The chemical constituents of cells include proteins – some in colloidal solution in the cytoplasm, some particulate in the organelles – droplets of fats (lipids), and carbohydrates as simple soluble compounds or granules of complex polysaccharides. There are also *pigments* in cells, such as the red haemoglobin of blood, its pink myoglobin derivative in muscle, its green and red degradation products in bile and faeces, and the photosensitive visual purple of the retina.

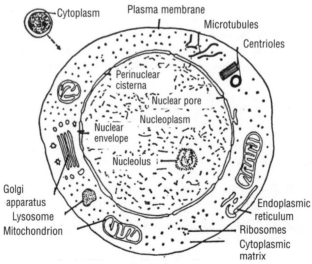

Figure 2.2 A diagram of a cell illustrating the structures that are essential for its survival (from *A Companion to Medical Studies*, Vol. 1)

The cells of different tissues may be greatly modified for their functions: the red blood cell has lost its nucleus, the nerve cell may have an enormously elongated axon process to carry stimuli, and the striated muscle cell can contract in length. Despite these modifications, the essential components remain recognizable in each case.

DNA

Every cell in the human body (except germ cells) carries within its nucleus the 46 chromosomes in 23 pairs that are characteristic of *Homo sapiens*. There are 22 pairs, the *autosomes*, not involved in sex determination and one pair of *sex chromosomes*. They are not clearly visible until the preliminary phase of cell division. Each DNA molecule contained in the chromosomes consists of a long pair of spirally wound strands – the double helix. This consists of a backbone of sugar (deoxyribose) and phosphate molecules with interposed purine or pyrimidine bases – adenine (A), thymine (T), guanine (G) and cytosine (C). The bases on one strand face those of the other in precise and regularly repeated fashion, like the rungs of a twisted ladder, and the order of these base pairs is:

▶ A and T

▶ C and G.

These contain the information or *genetic code* controlling protein synthesis, ensuring that the constituent amino acids are inserted in the proper places in the growing polypeptide chains (see Chapter 12).

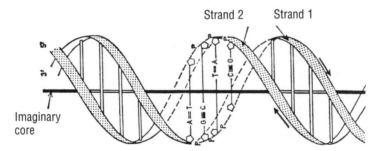

Figure 2.3 DNA thread of two strands round an imaginary core in an anti-parallel manner; A, T, G and C represent complementary base parts (from *A Companion to Medical Studies*, Vol. 1)

The organization is complex and is effected via 'messenger' and 'transfer' RNA molecules.

In RNA the sugar is ribose, thymine (T) is replaced by uracil, and the molecule is a single strand, roughly half that of a DNA molecule, with a row of single bases attached. The DNA of the cell is not all confined to the nucleus. Some is in the cytoplasm, concerned wth protein synthesis, but that of the nucleus is contained in the separate chromosomes.

A human DNA molecule is about a metre long and a twenty-millionth of a metre wide. The human *genome* is the total amount of DNA spread over the chromosomes of the cell, and it contains some three billion 'letters' of genetic code.

A *gene* is a segment of a DNA molecule that programmes the formation of a particular protein. The smallest gene has a sequence of about two thousand base pairs, but most genes are much larger. The total amount of information they contain, about 30,000–100,000 genes in humans, is the genome and each species of living creature has its own genome. No one genome is quite identical to another. Genes are separated by long stretches of DNA that seem to be meaningless – 'junk' DNA – with no discernible function. The genes control all structure and function and the transmission of inherited characteristics. In June 2000, the read-out of virtually the entire human genome was published, but much work remains to be done before this can be of any practical use.

> **Nugget**
>
> It would take 9.5 years of non-stop reading out loud to read a person's genome base by base.

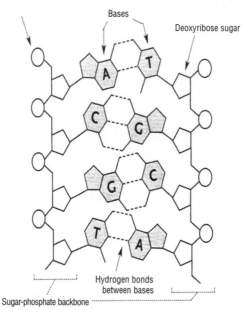

Figure 2.4 Chemical structure of DNA

In this diagram a short stretch of the familar double helix, formed by twin sugar-phosphate backbones, has been untwisted to reveal the molecule's ladder-like structure of the four bases, adenine (A), thymine (T), cytosine (C) and guanine (G) (from *Here Be Dragons* by David Koerner and Simon LeVay 1999 Oxford University Press Inc., by permission)

CELL DIVISION (MITOSIS)

Cell division in humans is an elaborate process of *mitosis*, evolved to ensure duplication of the original cell and exact sharing of its genetic material (Figure 2.6). Most of the body cells divide to allow growth and repair, but the neurones of the central nervous system cannot do so.

1 In the resting *interphase*, the chromosomes are not visible as separate structures.

2 As mitosis approaches, they become visible as threads within the nucleus (*prophase*).

3 The centrioles divide and move to opposite poles of the cell, as *asters* from which radiates a controlling *spindle* of protein fibres.

4 The nuclear membrane disappears. The chromosomes are now seen to be double and align themselves in the plane of the spindle between the asters.

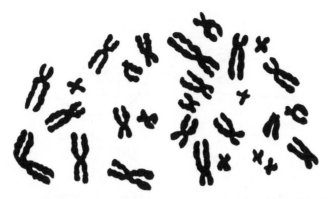

Figure 2.5 Chromosomes of a normal human male cell from a culture of peripheral blood (adapted from *A Companion to Medical Studies*, Vol. 1)

5 The two halves of each chromosome separate and are pulled by contracting spindle fibres to opposite ends of the cell to form the new chromosomes of the two daughter nuclei.

6 The cytoplasm of the cell cleaves in two and new nuclear membranes form around each separated bundle of chromosomes, reconstituting two daughter cells and two daughter nuclei genetically identical to the parent nucleus.

Because the body cells as a whole (*somatic cells*) have paired chromosomes, they are called *diploid*, and mitosis ensures that they remain so. But in the formation of the *germ cells* (ovum and spermatozoön) a different mechanism is involved (see Chapter 22).

Tissues

The body is composed of different *tissues*, each of which is an aggregation of similar cells, together with an intercellular ground substance or *matrix*.

There are only four essential tissues:

▶ *epithelium*

▶ *connective*

▶ *muscular*

▶ *nervous*.

These tissues may be modified in various ways but together they compose the structure of the body. All organs consist of at least two tissue types, usually more. They differ in the nature of their component cells and the matrix.

All tissues, however specialized, are repaired after injury by a fibrous connective tissue scar. The only exception, where the original structure is reproduced, is bone.

CONNECTIVE TISSUE

This is the framework of the body. It is characterized by the large amount of matrix, and the exact nature of a connective tissue depends on its matrix and the cells and fibres it contains. It includes widely different forms: bone, cartilage, fibrous tissue, elastic tissue, blood and lymph.

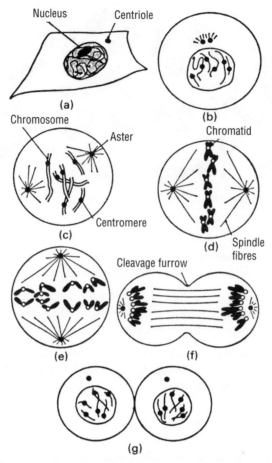

Figure 2.6 Mitosis: (a) interphase (46 chromosomes); (b) early prophase: centriole divides, chromosomes appear as threads; (c) late prophase: centrioles separate, aster and spindles form, chromosomes seen to be double, nuclear membrane disappears; (d) metaphase: chromosomes line up in equatorial plane; (e) early anaphase: centromeres divide, chromatids begin to separate; (f) late anaphase: separation of chromatids continues, cleavage furrow forms; (g) telophase: two daughter cells each with 46 chromosomes (from *A Companion to Medical Studies*, Vol. 1)

In simple *areolar* tissue there is a semi-fluid matrix containing mucopolysaccharides in which run strands of protein fibres – white (unyielding) and yellow (elastic) – together with rather sparse cells, an arrangement allowing gliding and stretching. This tissue forms the loose packing in clefts and spaces of the body, such as in the subcutaneous layer between skin and deep structures, and the interstitial framework of individual organs. This tissue contains much of the *extracellular fluid* of the body.

An excess of white fibres is found in the firm fibrous tissue of tendons and ligaments and sheets of membrane (*fasciae*). Elastic tissues predominate where stretch and recoil are important, as in the arteries. *Adipose* tissue is a store of triglycerides and a source of energy. It insulates against cold and buffers the internal organs, and is more developed in women. Most of the adipose tissue is located under the skin and may disappear almost entirely in starvation. It is divided into lobules by fibrous septa and is closely packed with cells, each of which is like a signet ring with its nucleus at one side and a distending fat globule.

In *blood* and *lymph* tissues the matrix is fluid, while the cells are modified to carry dissolved gases or deal with invading microorganisms. In *cartilage* and *bone* tissue the matrix has been hardened by impregnation with mineral salts.

In general, the cells of connective tissue fall into four types:

▶ *fibroblasts*, which manufacture fibres

▶ *mast cells*, concerned with carbohydrate metabolism, which are sentinels against foreign invasion and contain histamine to excite an inflammatory response

▶ *plasma cells*, which form antibodies against bacteria

▶ *phagocytes*, which engulf and devour foreign material and the debris of dead cells, and may be mobile.

Mast cells, plasma cells and phagocytes all enter connective tissue from the blood or lymph.

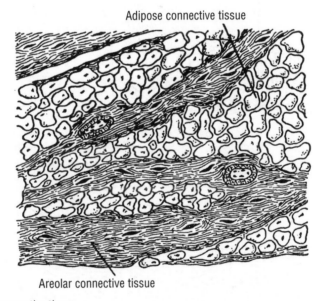

Figure 2.7 A simple connective tissue

EPITHELIUM

Epithelia are sheets of cells covering body surfaces: the outer aspect of the skin and the internal surfaces lining the body cavities. In addition they line the vessels, the respiratory, digestive and urinary tracts, and the glands opening into them. Both internal and external surfaces of the body possess this intact epithelial envelope as a barrier between the outside world (including, paradoxically, the contents of the bowel and hollow organs) and the deeper intermediate structures. These inner and outer linings blend at the orifices of the various cavities – at the lips, anus, nose and urethral apertures, where the moist linings or *mucous membranes* of the digestive, respiratory and urinary tracts become continuous with the skin.

Epithelium is very cellular and simple in arrangement, with little ground substance. The simplest form is one cell thick, with a basement membrane. It may be *squamous*, arranged in mosaic or pavement fashion, as at the skin surface; *columnar*, with tall regimented cells; or *transitional* as in the bladder mucosa, where it stretches to allow distension. Any form may be *stratified* in row upon row of cells. The epithelium of internal organs, especially the bowel, may have *villi,* finger-like processes which greatly increase the area available for absorption. At the skin the outermost layer becomes cornified. In some mucous membranes, as in the airways, the outer columnar cells may bear mobile hair-like processes or *cilia,* which waft mucus to the exterior.

It is epithelium that gives rise to the hair, nails and teeth; *endothelium* is the name given to the smooth linings of blood vessels; the glistening layers that line the great body cavities, pleural and peritoneal, are called *mesothelia*; and a *carcinoma* is a form of cancer in which epithelial cells have lost their adhesions and invaded adjacent tissues. Malignant tumours arising from connective tissue in various organs are called *sarcomas*.

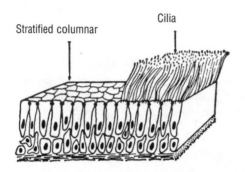

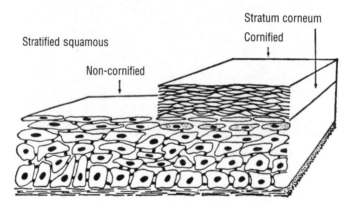

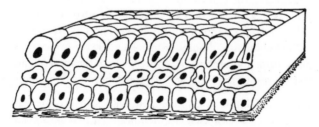

Figure 2.8 The stratified epithelia (from *A Companion to Medical Studies*, Vol. 1)

GLANDS

These consist of simple or complex arrangements of epithelial cells forming useful substances in secretory cells. The simplest form is the single *goblet cell* of the intestinal mucosa. But they may be more complex: tubular – simple, branched or coiled – or compound, with a branching duct system ending in blind pouches or *acini*. Acini are either *mucous* or *serous*. The cells of mucous acini appear empty; their clear cytoplasm contains polysaccharides. The serous acinar cells are granular, have large nuclei, and manufacture protein enzymes. Glands may also produce salt and solute (sweat glands), acid (gastric mucosa), alkali (duodenal mucosa) or grease (subcutaneous glands). They may secrete continuously, or only in response to some stimulus, such as the entry of food into the stomach. They may be independent, or under nervous, chemical or hormonal control.

Exocrine glands discharge their secretion at the skin surface or into an internal organ. *Endocrine* glands discharge directly into the bloodstream and their *hormones* affect remote target organs.

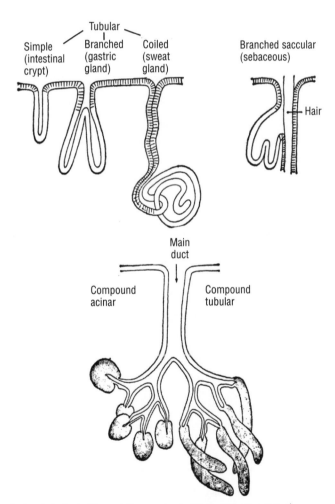

Figure 2.9 Types of glands (adapted from *A Companion to Medical Studies*, Vol. 1)

Test yourself

1 Which cell is visible to the naked eye?
   a   Spermatozoön
   b   Mitochondria
   c   Ovum
   d   White blood cell

2 Where is most of the ribonucleic acid (RNA) located in the cell?
   a   Nucleus
   b   Cytoplasm
   c   Cell membrane
   d   Golgi apparatus

3 Where are chromosomes located in the cell?
   a   Nucleus
   b   Cytoplasm
   c   Centrioles
   d   Endoplasmic reticulum

4 What type of cell does *not* contain a nucleus?
   a   White blood cell
   b   Red blood cell
   c   Nerve cell
   d   Muscle cell

5 Identify which base in DNA is paired with adenine (A).
   a   Thymine
   b   Guanine
   c   Cytosine
   d   Uracil

6 What stage of mitosis follows interphase?
   a   Early prophase
   b   Late prophase
   c   Metaphase
   d   Anaphase

7 Which of the following is *not* an essential type of tissue?
   a   Epithelium
   b   Connective
   c   Muscular
   d   Cardiac

8 What tissues are glands made of?
   a   Muscular
   b   Epithelial
   c   Nervous
   d   Connective

9 Which glands produce salt and solute?
    a   Sweat glands
    b   Gastric mucosa
    c   Duodenal mucosa
    d   Subcutaneous glands

10 Which of the following is *not* a type of epithelium tissue?
    a   Squamous
    b   Columnar
    c   Transitional
    d   Fibroblast

3

The skeleton

In this chapter you will learn about:

▶ *the role of cartilage in the body*

▶ *the different types of bones in the skeleton*

▶ *bone growth*

▶ *the different types of joints and movements permitted at each*

▶ *the roles of the bones in the axial and appendicular skeletons.*

Cartilage

Cartilage, with bone, is a major component of the skeleton. The greater part of the skeleton is originally laid down as cartilage during intrauterine life. It is found where rigidity and resilience are needed, i.e. at the joint surfaces, the front ends of the ribs, and the supporting framework of trachea, bronchi, nose and ears. The main types of cartilage are:

1 *hyaline articular* cartilage, smooth cartilage found at joint surfaces, with very low friction, giving a slippery surface

2 tough *fibrocartilage* containing white fibrous tissue, found in the intervertebral discs and the plates or *menisci* that project into certain joints, e.g. the semilunar cartilages of the knee

3 *elastic* cartilage, found in flapping structures like the external ear and epiglottis.

Cartilage is a gristly tissue, largely without blood supply, and it cannot repair itself after injury; instead it is replaced by fibrous scar tissue. It is a connective tissue whose matrix has become solidified as a firm gel bound together by fibrous proteins (*collagen* and *elastin*). Its organic content includes the complex molecule of chondromucoprotein linked with the polysaccharide chondroitin sulphate. The *chondroblast* or cartilage cell is plump and rounded, and distributed throughout the tissue in small groups. It is responsible for the formation of the matrix and is nourished by diffusion from the fluid of the adjacent joint, and to some extent from the underlying bone. Disuse leads to the degeneration of articular cartilage, whereas activity aids the diffusion of nutrients.

Cartilage has the property of *calcification*, i.e. calcium salts are deposited in it, rendering it tough and opaque. Calcification is a normal process during growth, as a preliminary to *ossification* of the cartilage precursors of the bones. But it is also an ageing process; in later life there may be fibrillary degeneration of joint cartilage, the wear and tear process of *osteoarthritis*.

Bone

Bone is a specialized connective tissue of mechanical function, though it has an important connection with mineral metabolism and houses the marrow in which new blood cells are formed.

Bone is the hardest tissue except for the teeth. It forms the framework of the body, protects the internal organs, gives attachments to the tendons and muscles, and constitutes the levers they move. Its structure combines strength with economy of material, its internal struts or *trabeculae* being disposed to meet maximum load. There is a considerable safety margin, e.g. the upper end of the femur can support a vertical load of a ton (900 kg) and bears three times the body weight at every step. Bone is subject to compression, tension, twisting and bending strains, and withstands these by its strength and *elasticity*. In old age and some diseases, this strength is impaired and fracture can occur.

Bone is a blend of:

▶ an *organic* fibrocellular matrix or *osteoid*

▶ a *mineral* matrix consisting of calcium phosphate and carbonate, magnesium and fluoride in crystalline form.

The organic component can be removed by burning, leaving a brittle mineral skeleton; the inorganic matter can be dissolved by acids, leaving decalcified flexible bone. The osteoid consists of collagen fibres, plus some ground substance containing mucopolysaccharides and protein. The mineral matrix has a rigid crystalline structure, known as *hydroxyapatite*, represented as $Ca_{10}(PO_4)_6(OH)_2$, wth variable amounts of other ions – magnesium, sodium, carbonate, citrate and fluoride. The collagen matrix acts as a site for the crystallization of minerals from the dissolved calcium and phosphate of the tissue fluids under the influence of an enzyme, phosphatase, that concentrates phosphate ions and initiates crystal growth. In health the whole of the bone substance is mineralized, and unmineralized osteoid is only found where bone is being formed rapidly, e.g. at fracture sites. In certain diseases, such as rickets, mineralization is defective, the bone is soft and yielding, and deformities occur. Sometimes, as in endemic fluorosis, mineralization is excessive and the bone is hard but brittle.

The salts of bone are in dynamic interchange with the ions of the body fluids, and this is the basis of the constant process of remodelling and replacement of bone substance. The calcium of the blood is in equilibrium with that of bone in a constant interchange to maintain the blood calcium within the narrow limits of 9–11 mg/100 ml, a process influenced by the secretions of the thyroid and parathyroid glands (Chapter 21), and by vitamin D and the local pH.

Bone contains 99% of the total calcium of the body, 88% of the phosphate, 80% of the carbonate, 70% of the citrate and 50% of the magnesium. It is constantly subject to simultaneous formation and reabsorption, more so during growth, with bone formation in the ascendant. During adult and early middle life the *bone mass* remains fairly constant in health, but in later life reabsorption predominates and the bone mass decreases so that it appears more translucent in X-rays – this process is known as *osteoporosis*. This begins earlier in women and accounts for the frequency of hip fractures in older women.

Microscopically, the dense outer cortical bone of the shafts of long bones is arranged in layers, or *lamellae*, containing small clefts, the *lacunae*, occupied by the bone cells, *osteoblasts*, whose processes ramify around. The lamellae are arranged around a central *Haversian canal* containing blood vessels and nerves, and a bone consists of a great many such Haversian systems, or *osteones*. This is modified near the surface of the shaft, where the lamellae run parallel and there are no canals. The *osteoblasts* are specialized connective tissue cells that form the organic matrix and contain alkaline phosphatase which catalyses bone formation.

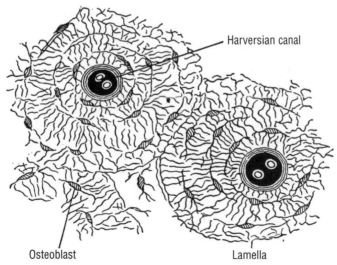

Figure 3.1 Microscopic structure of bone

Wherever bone is being absorbed or remodelled, there is another type of cell, the *osteoclast*, a giant multinucleate cell that actively engulfs bone substance. The two types of cell work in concert to shape the internal architecture of bone in response to mechanical stress.

The *periosteum* is a tough ensheathing membrane surrounding bone, except where it is covered by articular cartilage. Its outer fibrous layer is mainly supporting; it carries part of the bone's blood supply and gives attachment to muscles and ligaments. From it, fibres penetrate into the bone. In the deeper layer of the periosteum are osteoblasts responsible for increase in girth of the bone during growth. In children the periosteum is easily peeled off the underlying bone, especially in injury or infection. The attachment is much firmer in adult life.

TYPES OF BONE

There are five main types of bone.

1 *Long bones* are those of the limbs, e.g. the humerus in the arm, the femur in the thigh. They have a *shaft* which is roughly cylindrical, but sometimes polygonal or triangular in section, and two expanded ends, sometimes rounded off as a *head* or widened into *condyles*. The bone ends take part in the adjacent joints and are covered wth smooth articular cartilage to facilitate movement; the two ends forming a joint are enclosed in a common joint *capsule*. The periosteum covering the shaft becomes continuous with the capsule at the end of the bone.

   Cross-section of a long bone shows the outer *cortex* of bony substance and the central *medullary cavity.* The cortex has an outer portion of dense compact bone surrounding a meshwork of spongy or *cancellous* bone; at the ends, where the medullary cavity stops, there is a solid mass of spongy bone (Figure 3.4).

   Fatty *yellow marrow* occupies the medulla in the adult and the interstices of the spongy bone are filled with *red marrow*, responsible for formation of red and white blood cells. At birth and in the early years, red marrow occupies the whole shaft but retreats to the ends with growth. This may be reversed later if anaemia or haemorrhage makes extra demands on blood formation. In chronic anaemias, the shafts are occupied with red marrow working overtime.

   In adult life, active red marrow is found mainly in the flat bones, skull, ribs, sternum (breastbone) and pelvis, and also in the vertebrae of the spinal column.

2 *Short bones* are squat, cuboidal or irregular, composed entirely of spongy bone with red marrow and a thin compact shell. They occur at the wrist (*carpus*), the corresponding part of the foot (*tarsus*), and the vertebrae.

3 *Flat bones* include the skull vault, ribs and scapula (shoulder-blade). They consist of two plates of compact bone sandwiching a thin spongy layer; in the skull these layers are called the inner and outer *table*s, with the *diploë* between. Certain bones of the skull are expanded by air-containing cavities replacing this spongy layer – the *air sinuses*.

4 *Irregular bones* include the vertebrae and facila bones. They have complex individual shapes and perform a variety of functions, including protection and muscle attachment.

5 *Sesamoid bones* are tiny, rounded, pebbly masses found within certain tendons at points of friction adjacent to joints; by far the largest is the *patella* (knee-cap).

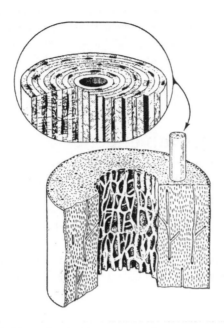

Figure 3.2 Diagram of cortical bone showing the structure of an osteone

These rod-shaped units of bone structure are made up of a series of lamellae (from *Electrical Effects in Bone* by C. Andrew L. Bassett. Copyright 1965 by Scientific American Inc. All rights reserved)

BLOOD SUPPLY OF BONE

A typical long bone has four sets of vessels (Figure 3.3). The *epiphyseal* vessels supply the centres of ossification at the growing ends. There is a separate set of *metaphyseal* vessels,

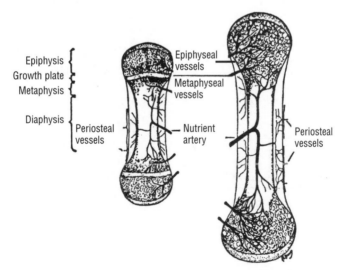

Figure 3.3 Diagram showing the arrangement of blood vessels in bone before and after the growth period

Note that the growth plate isolates the epiphyseal from the metaphyseal circulation during the growing period, but that they anastomose freely after growth has ceased The periosteal vessels supply the osteogenic layer on the bone surface during the growth period, but penetrate the cortex and supplement the nutrient artery in adult life (from *A Companion to Medical Science*, Vol. 1)

which are important in connection with growth. The shaft is supplied by one or more large *nutrient arteries* which penetrate the cortex and enter the marrow cavity. Finally, the *periosteal* vessels nourish the outer layers of the shaft. If the nutrient artery is damaged, or if the periosteum is stripped off by infection or injury, much of the shaft may die. The bone surface is dotted with numerous, tiny apertures (*foramina*) for blood vessels, with a larger *nutrient foramen* for the main artery near the middle of the shaft.

Development and growth of bone

In the embryo of a few weeks the central core of the primitive gelatinous connective tissue (*mesenchyme*) of the limb buds becomes transformed into an axial rod of cartilage. This rod is absorbed at the sites of the future joints, e.g. the elbow and knee, demarcating arm from forearm and thigh from leg, and split longitudinally in the distal segment as the forerunners of the paired radius and ulna, or tibia and fibula. Most of the skeleton is well formed in cartilage by the sixth week, and at the seventh week a *centre of ossification* develops in the midshaft of each long bone. Bone cells appear, the matrix is impregnated with calcium salts, and ossification spreads up and down the shaft until, at birth, the long bones are entirely ossified except for their cartilaginous ends.

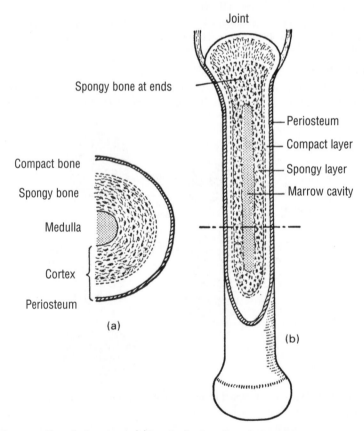

Figure 3.4 (a) Cross-section of a long bone; (b) longitudinal section of a long bone

EPIPHYSES

There now develops an arrangement, peculiar to mammals, allowing continuous growth in length (or height) during the years preceding maturity. In the first years of life, separate *secondary centres* of ossification appear in the bone ends, which ossify completely except for a thin cartilage *growth-plate* separating them from the shaft. The rounded end is called the *epiphysis*. The epiphyseal plate consists of longitudinal columns of cartilage cells reproducing themselves continuously on the shaft side of the plate and continuously turning into new bone, which is pushed away into the shaft, enabling it to grow in length. Meanwhile, the plate retains its integrity, until the epiphysis fuses with the shaft by ossifying across the intervening cartilage, at the age of 18–20 in men, 16–18 in women (see Figure 3.5).

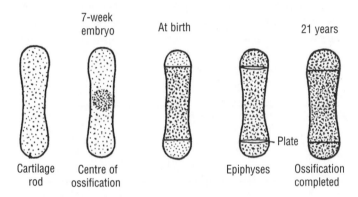

Figure 3.5 Stages in the ossification of a long bone

Although each end of a long bone is a growing end, one epiphysis appears earlier than the other and fuses later, and is more responsible for growth in length. In the arm these principal growing points are at the shoulder end of the humerus and the wrist end of the forearm bones; in the lower limb at the knee end of all three bones. The main nutrient artery of the shaft enters the bone at an angle directed away from the faster growing end. Complete destruction of an epiphysis causes cessation of growth and stunting; partial destruction causes a distortion of growth in which the limb turns away from the still active side.

A few bones, notably the skull vault, mandible and clavicle, are not preformed in cartilage, but are ossified directly in primitive membranous tissue and have no epiphyses.

▶ Remodelling

The articular cartilage, or joint surface, which covers the free end of the epiphysis, persists throughout life. It is that portion of the original cartilage system that has not been replaced by bone formed in the secondary centre of ossification. The shaft of the long bone grows in thickness by the apposition of new bone within the periosteum. As bone is added externally, it is removed from the cortex internally; so the shaft remains tubular and strong without being massive, and allows room for the marrow. This process of external deposition and internal reabsorption is called *remodelling*. Remodelling also occurs at the *metaphysis*, the 'neck' of the bone just on the shaft side of the epiphyseal plate.

The hard bones of the dried skeleton obscure their *plasticity* during life, for they respond to stresses much as a tree responds to the prevailing wind. The more work put on them the more they hypertrophy, particularly in response to compression or muscle pull, and the internal arrangement of bone trabeculae is a mechanical solution to the engineering problems imposed. Electrical forces are also involved; bone formation can be promoted by electrical induction and any sudden deformation of bone produces an electric current.

GROWTH

The foetus grows continuously, but its rate of growth varies considerably. It is slow during the first two months, reaches a peak at four to five months, and slows again in late pregnancy. The rate falls rapidly after birth, and then more slowly until puberty, when a *growth spurt* occurs. All growth in height ceases at around eighteen years in boys and sixteen years in girls, because of epiphyseal fusion with the shafts of the long ones. The changing proportions of the different components of the body are shown in Figure 3.6. In early foetal life, the head is disproportionately large because of the precocious development of the brain; at maturity, the lower limbs make the major contribution. Before puberty, the legs grow faster than the trunk and boys are generally taller than girls because they have a longer prepubertal period. Most of the adolescent increment occurs in the trunk. In old age height decreases, due to bending and degenerative changes and collapse in the spine.

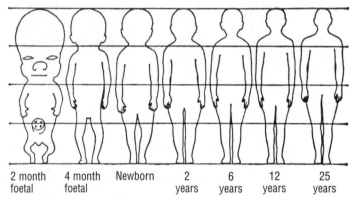

| 2 month foetal | 4 month foetal | Newborn | 2 years | 6 years | 12 years | 25 years |

Figure 3.6 To show the changing proportions of the body during growth and development (from *A Companion to Medical Studies*, Vol. 1)

Nugget

The longest bone in the human body is the femur; the smallest bone, located in the middle ear, is called the staples and is only 2.8 millimetres long.

▶ Factors controlling bone growth

Growth in height depends on the growth in length of the long bones. This in turn depends on the growth in the epiphyseal cartilage plates, so growth rate and final stature are determined by the rate of proliferation of cartilage cells in the growth plate and its active duration. These are affected by a number of factors:

▶ *genetic* influence, e.g. identical twins are of similar height and tall parents tend to have tall children

▶ *malnutrition* in childhood stunts growth

▶ *endocrine* glands affect bone growth.

The growth hormone (*somatotrophin*) secreted by the anterior lobe of the *pituitary* gland promotes growth from birth to adolescence by stimulating the reproduction of cartilage cells. An excess of secretion in childhood causes *gigantism,* the sufferer reaching seven or eight feet in height. Its action in the adult is less dramatic, but there is hypertrophy of the skeleton and soft tissues, especially of the hands, feet, face and tongue – the condition known as *acromegaly.* Deficiency of growth hormone in early childhood is one cause of *dwarfism.* The thyroid hormones are also important. In congenital deficiency (*cretinism*), there is stunting of physical and mental growth, and the epiphyses are late in appearing and unfused in middle age. *Sex hormones* secreted by the sex organs and adrenal glands cause the adolescent growth spurt.

Nugget

Bones stop growing in length during puberty; however, bone density and strength change over the course of a person's life.

Joints and movements

TYPES OF JOINT

The articulations between the bones vary in mobility in different parts of the skeleton. Some are very mobile, some allow no movement. There are three main types.

1   *Fibrous.* As a zigzag *suture* between the skull bones, whose interlocking irregularities are joined by a narrow fibrous strand, usually obliterated by ossification in later life; or as a *syndesmosis*, which allows some stretching and twisting, as in the distal tibiofibular joint; or as a broad sheet or *interosseous membrane* between the shafts of paired bones radius and ulna, tibia and fibula.

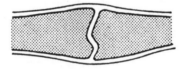

Figure 3.7 Suture

2   *Cartilaginous.* Here the cartilage coverings of the bone ends are united by a pad of thick fibrocartilage, e.g. the cushion between the two halves of the pelvis at the pubic symphysis (Figure 3.12), or interposed as intervertebral discs between the vertebral bodies. Movement of such a joint is very limited, but the total range over a series of vertebrae may be considerable.

3   *Synovial.* These allow free movement. The bone ends are capped with smooth hyaline *articular cartilage*, and the joint cavity is enclosed within the sleeve formed by the fibrous joint capsule, stretching from one bone to the other and continuous with the periosteum. *Ligaments* made of strong fibrous tissue hold the bone ends together. Foremost of these is the main capsular ligament itself, and in addition there may be localized bands within the substance of the capsule, and accessory ligaments traversing the joint or outside it altogether. Tendons arising within a joint function as accessory ligaments, e.g. the long head of the biceps at the shoulder joint. In some cases complete or incomplete cartilage plates, or *menisci*, project into the joint cavity as partitions; these occur at either end of the clavicle,

between the lower jaw and the skull (*temporomandibular joint*) and at the knee (the *semilunar cartilages* often torn in sports); also at the wrist at the inferior radio-ulnar joint.

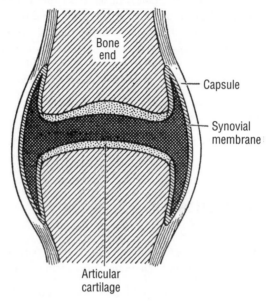

Figure 3.8 A synovial joint

Normally, the joint space is only potential, as the cartilage and soft tissue within lie everywhere in contact under a negative pressure; it becomes a reality only when the capsule is distended by fluid or air is admitted. The capsule is lined by a smooth *synovial membrane*, which is reflected onto the bones but disappears at the periphery of the cartilage surfaces. This synovial membrane packs the joint and secretes the lubricating *synovial fluid*. This is a yellowish, viscid fluid secreted from the blood, which also nourishes the cartilage; the white cells it contains scavenge the debris from the moving surfaces. Where the bones do not fit together accurately there may be fatty pads between the capsule and synovial membrane which bulge into the dead space and cushion movement. Any structure traversing the joint, such as a ligament or tendon, carries an investing *sheath* of synovial membrane.

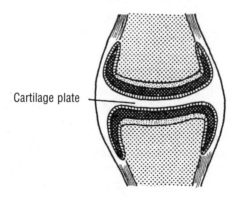

Figure 3.9 A synovial joint partitioned by a cartilage plate

STABILITY

In all joints there has to be a compromise between stability and mobility. The former is mainly due to factors the most important of which are the strength of the ligaments and the tone of surrounding muscles. Great mobility is essential at the shoulder, so the capsule is lax and the large humeral head fits poorly into the shallow socket of the scapula; but this inherent instability is countered by the tendons of the small rotator muscles blending with the capsule to compensate for the anatomic deficiency. In the hip, however, stability is all-important; the femoral head is deeply embedded in the socket of the acetabulum and the adjacent muscles do not have to blend with the capsule. Minor supportive factors are interlocking of the bone ends, not often very secure, and the cohesion produced by atmospheric pressure.

The joint surfaces are least closely fitting during the middle range of movement and *close-packed* at the extremes of range, when the joint resists stresses with minimal muscle guarding. There is also a *rest position* when it is not in use. The ligaments are important reinforcements, but have little stretch and are not intended to resist severe strain on their own or for long. This is the task of the surrounding muscles, which never completely relax during activity but pay out slack through a resisted movement as their antagonists contract, in a manner impossible for the ligaments.

MOVEMENTS PERMITTED AT SYNOVIAL JOINTS

These are classified by the shape of the bone ends and the range of possible movements.

▶ The *plane joint* between the small flat adjacent surfaces of the carpal and tarsal bones allow minor gliding motion only.

▶ The *saddle* joint is self-explanatory, with reciprocating surfaces, the best example bring between the thumb metacarpal and the corresponding carpal bone (Figure 6.7).

▶ The *hinge* joint is common, as at the elbow, ankle and fingers; a convex surface is gripped within a concavity and strong collateral ligaments on each side allow motion round a transverse axis only. The capsule obviously has to be lax in front and behind to allow full flexion and extension.

▶ The *pivot* joint has a cylindrical bone end rotating on its long axis like a door hinge in its socket, e.g. the head of the radius at the elbow in rotation of the forearm, and the odontoid process of the axis within the ring of the atlas in the neck.

▶ The *ball-and-socket* joint of shoulder and hip has the greatest range, with one spherical and one concave surface moving on the other round an infinite number of axes. Its main movements (Figures 3.10, 3.11) are *flexion* and *extension* in an anteroposterior plane;

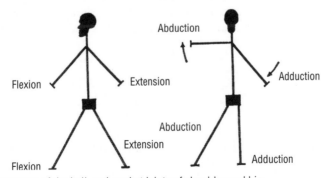

Figure 3.10 Movements of the ball-and-socket joints of shoulder and hip

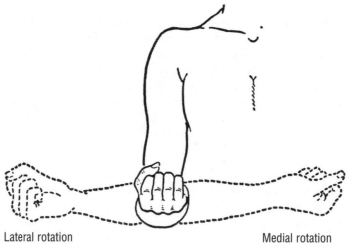

Lateral rotation   Medial rotation

Figure 3.11 Rotation at the shoulder

abduction and *adduction* in a mediolateral plane; *medial* and *lateral rotation* in the long axis of the limb; and *circumduction*, in which the limb describes the boundary of a cone in a combination of all the above.

The movements of joints are rarely as extensive as the shape of the bone ends would seem to allow, i.e. actual locking is only exceptionally a limiting factor. The soft parts are usually responsible for limiting the range, as in contact of the arm and forearm in bending the elbow and tension of the hamstrings in limiting flexion of the hip.

Axial and appendicular skeleton

The skeleton is divided into the central *axial* portion of the head and trunk, and the peripheral *appendicular* portion of the limb bones.

The *axial skeleton* consists of the skull, mandible (lower jaw), spine, sternum (breastbone) and twelve pairs of ribs. In addition, there are the small hyoid bone in the upper part of the neck between the larynx and the floor of the mouth and three tiny ossicles in each middle ear cavity.

SPINAL COLUMN

The spine is the bony axis of the body and consists of individual *vertebrae* in a segmental pattern. These articulate with each other, so that the sum of the limited motion between individual pairs is considerable. The column is traversed by a central canal, which encloses and protects the spinal cord, and it supports the weight of the trunk and transmits it to the lower limbs.

▶ **The vertebrae**

The vertebrae are grouped regionally as:

Cervical (neck)	7	*Sacral* (hip)	5
Thoracic (chest)	12	*Coccygeal* (tail)	4
Lumbar (abdominal)	5		

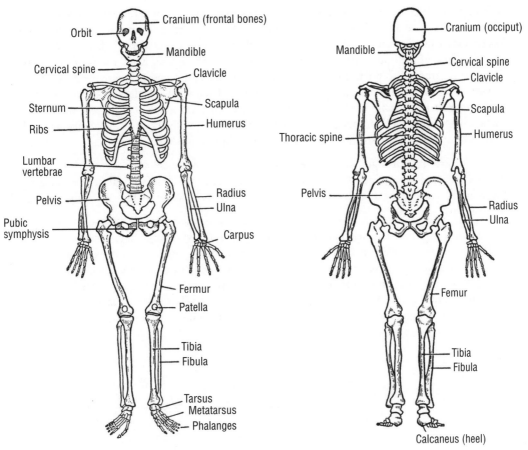

Figure 3.12 The skeleton, anterior view

Figure 3.13 The skeleton, posterior view

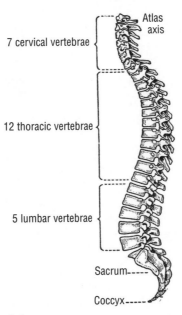

Figure 3.14 The spinal column, lateral view

Although modified in the different regions, the essential features of a vertebra are similar everywhere (Figure 3.15). In front a massive rounded *body* projects forward, with upper and lower surfaces facing the vertebrae above and below. Behind is the *neural arch*, which, along with the back of the body, forms a complete bony ring – the *neural canal* for the spinal cord. The arch is attached to the body by two pillars or *pedicles*; projecting at either side are the two *transverse processes*, and the *spinous process* projects backward and downward from the back of the arch under the skin of the back. There is a pair of small *articular facets* on the upper surface of the arch and another pair below for articulation with those of the adjacent vertebra. The vertebral body consists of spongy bone with a thin outer shell.

A pair of vertebrae fit together to leave on each side an *intervertebral foramen* for the exit of a spinal nerve. There are 34 pairs of spinal nerves since the first cervical nerve emerges between the base of the skull and the first cervical vertebra, while there is only one coccygeal nerve below.

The *cervical vertebrae* are delicate, with small bodies and a very large neural canal, as the cord is thickest at its upper end before it has given off most of its roots. The transverse processes are pierced by a canal transmitting the *vertebral artery*, which runs up the neck from the subclavian artery and enters the foramen magnum at the skull base to supply the hindbrain. The spinous process is short and bifid. The first cervical vertebra is the *atlas*, supporting the head. It is a simple bony ring without a body, and its upper surface articulates with the

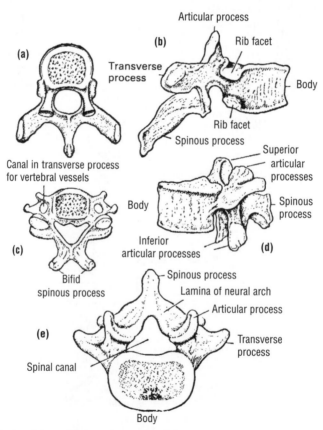

Figure 3.15 Some typical vertebrae: (a) thoracic vertebra; (b) thoracic vertebra, side view; (c) cervical vertebra; (d) lumbar vertebra, side view; (e) lumbar vertebra, from above

condyles of the occipital part of the skull to form the joint at which nodding occurs. The second cervical vertebra is the *axis*; from its upper surface a peg-like process, the *odontoid*, projects upward within the atlas ring. Skull and atlas rotate on this peg to turn the head from side to side.

The *thoracic vertebrae* increase in size downward. The bodies are heart-shaped, with a *facet* at each side for articulation with the heads of the ribs. A similar facet on the transverse process articulates with the neck of the rib. The neural canal is relatively small; the spinous processes project downward.

The *lumbar vertebrae* are under considerable load and have massive bodies. The neural canal is triangular and the squat spinous processes project horizontally backward.

The *sacral vertebrae* are fused into one bone, the *sacrum*, lodged between the innominate bones as the posterior segment of the pelvic ring. It is a large flattened triangular bone with anterior (pelvic) and posterior (subcutaneous) surfaces (Figure 3.16). The base articulates with the fifth lumbar vertebra at the lumbosacral joint, the apex with the coccyx below. It is placed obliquely, with its long axis directed backwards and the anterior surface is the hollow in which the rectum lies.

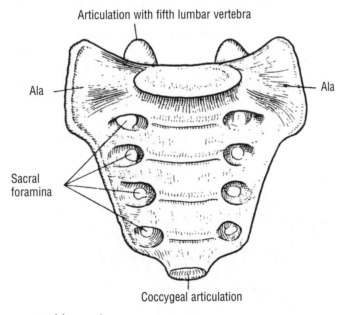

Articulation with fifth lumbar vertebra

Ala — — Ala

Sacral foramina

Coccygeal articulation

Figure 3.16 The sacrum, pelvic aspect

Although a single mass, the general features of the component vertebrae are still obvious, the bodies being indicated by transverse ridges. Four pairs of *foramina* transmit the sacral nerve roots, and the body of the bone is traversed by the *sacral canal*, which, at this level, contains not the spinal cord (this has ended in the upper lumbar region) but a thick bundle of nerve roots, the *cauda equina*.

Above, the *promontory* of the sacrum is a prominent ridge on the anterior surface which forms the back of the pelvic brim. On each side a wing or *ala* reaches out to form with the innominate bone the *sacroiliac joint*, a large interlocking pair of rough surfaces with little or no motion. The back of the sacrum is the lowermost origin of the *sacrospinalis muscle*.

The *coccyx* is the rudimentary tail appendage, a small irregular mass of four fused vertebrae; it is triangular, angled forward at the *sacrococcygeal joint* and quite solid.

▶ The spinal column as a whole

The vertebral bodies are separated by thick cushions of fibrocartilage, the *intervertebral discs*. These increase in depth down the column until in the lumbar region they are nearly 13 mm (1/2 inch). Each consists of an outer fibrous ring or *annulus* enclosing a gelatinous core under tension, the *nucleus pulposus*, which acts as a ball-bearing, The *intervertebral joint* consists of the disc between the bodies plus the simple plane synovial joints between the pairs of articular processes on the arches. Long strap-like ligaments bind the bodies in front and behind, the *anterior* and *posterior longitudinal ligaments* which traverse the length of the column; shorter bands connect the spinous and transverse processes, the whole being a flexible entity. The discs may rupture or prolapse backwards into the spinal canal. This occurs mainly in the lumbar region, where the extruded material presses on a sacral nerve root, causing sciatica. Extrusion higher, in the thoracic region, is rare but more serious as it endangers the spinal cord itself.

Although the spine is quite straight looked at from in front or behind, there is in side view a series of curvatures alternating in direction. The cervical and lumbar curves are forward convexities or *lordoses*; the thoracic and sacral curves are backward convexities or *kyphoses*. This arrangement of arcs, spanned by the spinal muscles, is mechanically sounder than a more rigid straight column. The child's spine at birth does not have the cervical and lumbar curves, only the anterior concavity of the whole spine (the crouched foetal position) and of the sacrum. As the head is held up in the first six months, the cervical lordosis develops, and the lumbar curve results from sitting and standing in the second year. In the adult, the vertebral bodies are somewhat wedge-shaped to conform to the shape of the curves.

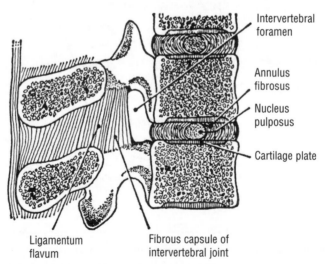

Figure 3.17 Median section of articulating thoracic vertebrae, lateral view (from *A Companion to Medical Studies*, Vol. 1)

In weakening bony diseases such as osteoporosis or tuberculosis, and in old age, there is a tendency for the thoracic kyphosis to become exaggerated and form an actual hump. The disease of *ankylosing spondylitis* fuses the joints and obliterates spinal movement, notably in the cervical region. A sideways curvature of the spine or *scoliosis*, usually in the thoracic

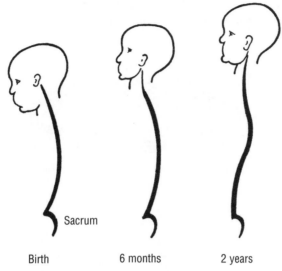

Sacrum

Birth   6 months   2 years

Figure 3.18 Spinal curves at different ages

or thoracolumbar region, is always abnormal. In young girls it is often simply postural and disappears on lying down. But it may be fixed, with a lateral convexity and some rotation of the vertebral bodies, creating a *kyphoscoliosis* or hump. This is of unknown origin and difficult to correct surgically.

▶ Spinal movements

These are:

1 *flexion,* forward bending, maximal in the cervical region, nil in the thoracic region, considerable in the lumbar spine

2 *extension,* backward bending, maximal in the lumbar and cervical regions

3 *rotation,* a twisting of the column on its longitudinal axis traversing the centres of the discs, freest in the upper thoracic region and negligible elsewhere

4 *lateral flexion* or sideways bending, maximal in lumbar and cervical regions

5 *circumduction,* a swaying movement combining all the above, in which the trunk describes the surface of a cone, apex below at the lumbosacral joint.

THE APPENDICULAR SKELETON

This part of the skeleton consists of the bones of girdles and bones of the limbs.

▶ Limb girdles

These are encircling arrangements of bones connecting the shoulder and hip regions to the axial skeleton, so as to provide a firm attachment for the corresponding limb while allowing a degree of mobility. The two girdles are very different, as the arm does not bear weight and must allow the freest use of the hand; the shoulder girdle is correspondingly unstable and its main central connections are muscular rather than bony. This is reversed at the hip, where stability is essential, and here the pelvic bones form a complete bony ring, firmly attached to the spine behind and to each other in front (Figure 3.19).

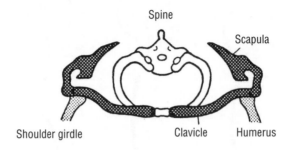

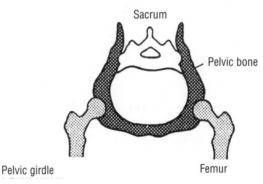

Figure 3.19 The limb girdles

Both girdles are based on a circular model, but the *shoulder girdle* is very wide and open behind, where the scapulae are only attached to the spine by muscles so that they can move freely on the trunk. Each half of the girdle is made up of two bones – scapula and clavicle – the latter a slender bone joining scapula to sternum and bearing part of the burden of the hanging arm. Halfway round the girdle on each side is the shallow glenoid fossa of the scapula, accommodating the humeral head at the shoulder joint.

In the *hip girdle* the bony ring is complete in front and behind in a tight circle. Each half is composed of a single pelvic (innominate) bone carrying a very deep acetabular socket for the head of the femur at the hip joint. Behind, the two pelvic bones are firmly attached to the spine – here the sacrum – at the sacroiliac joints. In front, they are firmly attached at the fibrocartilage of the pubic symphysis. The girdle is integrated with the axial skeleton and does not permit any accessory movements except those at the hip.

Test yourself

1 What type of cartilage is found at joint surfaces?
   a Fibrocartilage
   b Elastic
   c Hyaline
   d Elastin

2 Which mineral helps to ensure cartilage is tough?
   a Calcium
   b Magnesium
   c Zinc
   d Potassium

3 Identify one tissue in the body that is harder than bone.
   a Teeth
   b Nails
   c Hair
   d Ligaments

4 Which tissues catalyse bone formation?
   a Osteoclasts
   b Osteones
   c Osteoblasts
   d Lacunae

5 Which type of bones are composed entirely of spongy bone?
   a Long bones
   b Short bones
   c Flat bones
   d Sesamoid bones

6 Which bones are located within tendons?
   a Irregular bones
   b Short bones
   c Sesamoid bones
   d Flat bones

7 Which vessels supply the outer layers of the bone shaft?
   a Epiphyseal vessels
   b Metaphyseal vessels
   c Periosteal vessels
   d Nutrient foramen

8 Which of the following can stunt bone growth?
   a Malnutrition
   b Excess growth hormone
   c Genetic influences
   d Sex hormones

9 Which type of joint permits the greatest range of movement?
   a Fibrous
   b Cartilaginous
   c Synovial
   d Fixed

10 Which type of joint is the only one to permit circumduction movement?
   a Pivot
   b Ball-and-socket
   c Hinge
   d Saddle

4

Muscle

Virtually all movement in the human body is effected by contraction in muscle cells. The heartbeat, expulsion of urine, faeces or foetus, the rhythm of bowel movement, pupillary dilatation and contraction, are all examples of muscular activity.

Muscular tissue is composed of reddish fibres arranged in bundles. These fibres are highly specialized elongated cells marked by their ability to contract on stimulation. The muscles vary in the speed, strength and duration of their contractions. Some act through conscious intent, others function unconsciously. Some contract only if stimulated by nerve impulses, others have an intrinsic independent pattern of contraction, and yet others respond to circulating hormones. These varying functions are associated with different types of muscle structure.

1 *Smooth* muscle (also known as plain, unstriated, involuntary or visceral) is the most primitive. It forms the contractile coat of blood vessels and hollow internal organs like the bowel and bladder – structures that must work automatically, beyond conscious control, and regulated by the semi-independent autonomic or vegetative nervous system. The degree of control varies: some smooth muscle, such as that of the bladder, has an intrinsic rhythm which is only modifed by nervous stimuli. All smooth muscle is stimulated by being stretched. The typical smooth muscle cell is spindle-shaped, about 250 µm long, and has one large oval nucleus at its middle.

Nucleus

Figure 4.1 A smooth muscle fibre

2 *Skeletal* muscle (also known as striated, somatic or voluntary) has more intricate fibres which appear cross-striped under the microscope. They are found in the muscles attached to the skeleton and are under the conscious control of the central nervous system. These striped cells can be very long and may run for several centimetres from one attachment of the main muscle to the other. Each contains many hundreds of nuclei, placed to one side of

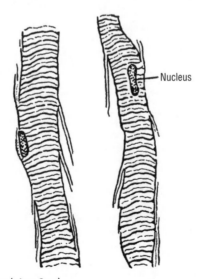

Nucleus

Figure 4.2 Skeletal muscle fibres (after *Gray*)

the cell, and each cell receives near its midpoint the termination of a nerve fibre from the brain or spinal cord. At this junction, the muscle cell and the nerve fibre form a complicated structure – the *myoneural junction* or *motor end-plate*.

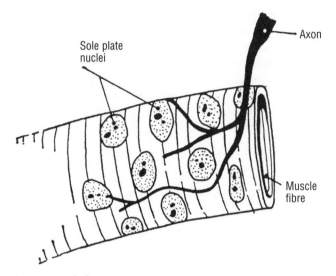

Figure 4.3 Drawing of a motor end-plate

Note the terminal arborization of the axon and the accumulation of nuclei in this region (from *A Companion to Medical Studies*, Vol. 1)

3 *Cardiac* muscle is found in the wall of the heart. Its fibres are striated, but not under voluntary control, and the cells are branched to form a network without clear demarcation between cells. There are less than a dozen nuclei to each cell, situated near its centre. Each cell has its own intrinsic rhythm.

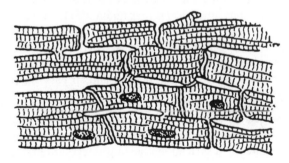

Figure 4.4 Cardiac muscle

The muscle cell

The muscle cell is specialized to convert chemical energy into contractile force, and is elongated along its axis of contraction. It is sheathed by an excitable membrane, the *sarcolemma*, while the cytoplasm is the *sarcoplasm*. Here the mitochondria are large and numerous, there are many glycogen granules, and a special feature is the presence of contractile protein *myofilaments*, which run the length of the cell and are only seen under the electron microscope. Where the filaments group together and become visible to light microscopy, they are known as *myofibrils*.

Myofilaments are of two types: thick and thin. The thin filaments consist of a protein, *actin*, in fibrillary form, which can also exist in globular form. The thick filaments consist of another protein, *myosin*. Each molecule has a rounded head and a long tail; the tails run together in thick strands and the heads project at the sides of the filaments. Both types of filaments lie side by side, and the contractile force is generated between opposing molecules, where actin globules and myosin heads come together like the row of a zip-fastener and the thin filaments slide past and overlap the thick ones. This is associated with a local increase in calcium ions brought about by a nerve impulse. As both filaments are polarized in the direction of their molecules, contraction takes place in one direction only, i.e. a muscle can actively contract but not actively lengthen.

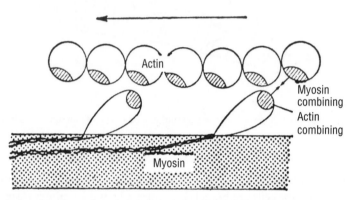

Figure 4.5 The interaction of myosin and actin (from *A Companion to Medical Studies*, Vol. 1)

In *smooth muscle* the filaments run along the long axis of the cell; they are not regularly patterned and no cross-striations are seen. There is no limit to the range of shortening and smooth muscle, though it contracts slowly, can exert great tension over long periods. This tension held for a long period of time is called *tonic* activity and is under the influence of the autonomic nervous system and circulating hormones. This type of muscle is important in the functioning of the heart, blood vessels, bowel, bladder and air passages.

In *skeletal muscle* there is a regular arrangement of the two types of filament, producing the typical cross-striation associated with the grouping of filaments into myofibrils that allows rapid contraction. A muscle fibre can shorten by over half its relaxed length, which is the degree of contraction required of a skeletal muscle to produce a full range of movement at the joint it controls.

Additional smooth muscle cells can be produced in response to requirements throughout adult life – as in the uterus in pregnancy. This is not possible in skeletal muscle, where any increase in bulk from exercise is due to increase in size, or *hypertrophy*, of individual cells. Muscles hypertrophy when stimulated by sex hormones at puberty, especially in males. They *atrophy*, or waste, with disuse; interruption of the motor nerve supply causes severe, rapid and permanent wasting.

We can distinguish between the striped muscles attached to the skeleton and concerned with rapid-response voluntary movements, controlled by the will through nervous messages from the brain, and the smooth muscle forming the walls of hollow internal organs and the blood vessels. The latter function without conscious attention, and are controlled by the semi-independent autonomic nervous system (see Chapter 20), but can continue to act to some extent even if denervated, for their nerves modulate rather than initiate activity.

Skeletal muscle

Voluntary muscles differ from involuntary by being able to contract in response to environmental changes. This response is dictated by the central nervous system and is called a *reflex* action. Figure 4.6 shows how a stimulus to the skin is conveyed by a sensory nerve fibre to the central nervous system and there relayed to a motor nerve cell whose fibre runs out to end in the muscle it stimulates to contract. This *reflex arc* is well established in very simple animals, and secures the immediate unthinking performance of protective actions such as blinking. In humans, reflex actions have come increasingly under the control of the will, i.e., the message is relayed up the spinal cord to the brain, which may modify the primitive response. It is like a ship being steered by a gyro-compass responding reflexly to any change in course, but providing information to the captain, who may intervene at any time.

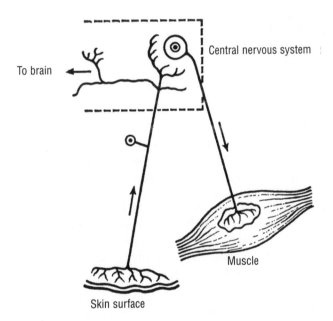

Figure 4.6 A simple reflex arc

The stimulus conveyed from the skin is relayed through a cell of the central nervous system to elicit an automatic contraction of the appropriate muscle

At the same time information is passed to the brain for conscious control of the proceedings (adapted from Starling's *Principles of Human Physiology*)

Every skeletal muscle not only has motor nerve fibres arriving from the central nervous system, but is also provided with sensory fibres conveying information on the degree of tension in the muscle. This *feedback* is important, as muscles act in groups and must be centrally coordinated.

Each muscle cell has one end-plate associated with the termination of a nerve fibre, and each motor nerve cell in the nucleus of a cranial or spinal motor nerve sends out one prolonged axon within that nerve, which branches in the muscle to supply a number of cells. Individual muscle cells can run the length of the muscle or end at intermediate levels. In the first case all the end-plates lie in a zone near the centre, in the second they are distributed throughout the muscle. The nerve cell and its axon exhibit 'all-or-none' behaviour; they transmit either an impulse of a certain magnitude or nothing at all. The arrival of an impulse at a motor end-plate causes a similar all-or-none depolarization of the related muscle cells, an action potential across the membrane (*sarcolemma*) of the muscle cell, and a transient rise in intracellular calcium that is the ultimate trigger for contraction. An individual fibre contracts by 50–60% of its relaxed length or not at all. But not all the fibres of a muscle necessarily contract simultaneously depending on the work to be done.

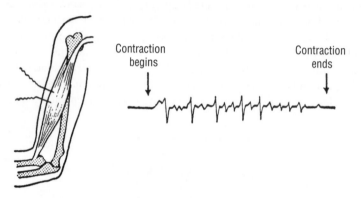

Figure 4.7 Electromyogram

MOTOR UNIT

The group of muscle cells supplied by a single neurone is known as a *motor unit*, and the number of cells it contains is related to the delicacy of the relevant movement, as there are fewer cells per unit where fine movements are concerned. To put it another way, the overall ratio of neurones to muscle fibres is much higher in the smaller muscles. In the muscles controlling movements of the eyes and fingers a single motor unit comprises perhaps only a dozen cells, whereas in the coarser muscles of the limbs and trunk there may be several hundred. An electrode placed over or in a muscle can detect the electrical activities of local motor units, and these can be amplified to give audible evidence of 'firing' – a machine-gun-like effect – or the visual record of an *electromyogram*.

The electrical impulse transmitted down the nerve fibre liberates acetylcholine at the end-plate and this increases the permeability of the muscle membrane to sodium, potassium and other cations. An electrical potential is generated which spreads through the muscle fibre at several metres a second and can be recorded. There is little electrical activity in a muscle at rest. Smooth muscle often has a double nerve supply; one group of fibres releases acetylcholine, and the other, noradrenaline. The first is excitatory and the other inhibitory; smooth muscle also responds to circulating hormones.

Muscle contraction

Muscles may be excited to contract by *stimuli* such as heat, pinching, chemical irritation or electrically. A constant galvanic current excites a single contraction at the start and end of its flow, at make and break; a faradic current does likewise, but the current is of much shorter duration and there is a minimum duration of flow below which there is no response. This *threshold* value is only a fraction of a second in warm-blooded animals. The response to electrical stimulation increases with the strength of current, as more fibres come into action. For each fibre, contraction is all-or-none; it contracts completely or not at all. After contraction there is a *refractory period* of a fraction of a second during which the muscle is inexcitable, a period of recovery and preparation for the next contraction.

If a skeletal muscle is repeatedly stimulated, it does not contract indefinitely but becomes *fatigued*; it does less work and eventually fails to respond. This is due to the using-up of all the sources of energy and the accumulation of wastes. In combating fatigue a good supply of blood and oxygen is essential. Fatigue in humans does not relate only to the muscle itself; it also involves tiring of the controlling nerve cells and fibres.

The generalized muscular contraction after death, *rigor mortis*, must not be confused with contraction in life. It is associated with the accumulation of wastes and irreversible changes in muscle proteins.

Nugget
Rigor mortis is accelerated by previous fatigue – the onset of it is therefore much more rapid in animals that have been chased and killed in a hunt.

MUSCLE CONTRACTION PRODUCING MOVEMENTS
A muscle moves the joint or joints between its origin and insertion (see above). Even when a movement is gravity-assisted, such as when the arm falls to the side, there is a controlled relaxation, which is really a continuous readjustment of the degree of contraction, i.e. 'relaxation' is an active process.

Ordinary contraction, when origin and insertion are allowed to approximate freely and the muscle belly shortens, is *isotonic*: the tension remains constant throughout. If the attachments are kept apart by resistance, the muscle cannot shorten and its tension rises – an *isometric* contraction of constant length. Most actions are performed against some degree of resistance, if only the weight of the part, and are a blend of isometric and isotonic.

TONUS
Living muscle is never completely relaxed except under deep anaesthesia or in some forms of meditation, but is aways in slight contraction or *tonus*. Tonus is essential to posture, especially the erect posture, as it holds the body against the strain of its own weight. In the feet and legs the ligaments would be subject to enormous stresses but for the guarding of the surrounding muscles. The spinal column is not a straight rod but, seen from the side, a series of curves whose general shape is maintained by the long muscles forming the cords of these arcs. *Postural reflexes* in the erect position aim to preserve an upright, forward-looking head and complex reciprocal actions of all the muscle groups in the calves, thighs, buttocks, spine and neck function to this end. Hence the necessity for a *sensory* nerve supply from muscles telegraphing the state of tension from moment to moment to the central nervous system.

Smooth muscle

Smooth muscle differs considerably from voluntary muscle in its inherent properties and nervous control. It has intrinsic rhythmic contractions and is guided by a different nervous system. The autonomic fibres to smooth muscle are the sympathetic and parasympathetic, stimulating it to relax and contract respectively and are mediated by different chemical agents. Smooth muscle is much more sluggish than voluntary muscle, it is very sensitive to stretching, which matters in hollow organs like the bowel and bladder which respond to distension or emptying. A *slowly* increasing distension, as when the bladder fills with urine, causes the muscle coat to relax proportionately, so that the fluid pressure does not rise until the last stage of distension, when the maximum pressure stimulates waves of rhythmic contraction which end in mass expulsion of the contents of the organ. *Rapid* distension provokes a more immediate response.

The entrance and exits to hollow organs are controlled by ring-shaped bands of smooth muscle, or *sphincters*, e.g. the cardiac and pyloric sphincters at each end of the stomach; these normally remain contracted until relaxation is needed to allow the organ to fill or empty. The nervous and chemical control of a sphincter must obviously be the converse of that of the organ as a whole, since it must contract while the organ relaxes and vice versa. Thus the sympathetic and parasympathetic nerves and their chemical agents have exactly opposite actions on the muscle wall of an organ and the sphincter muscles that guard its orifices.

Gross anatomy of muscles

The skeletal muscles, the flesh of the body, make up 40% of its weight and are divided into the *axial* muscles connected with trunk, head and neck, and the *appendicular* muscles of the limbs. The separate fibres are bound in bundles by connective tissue, and these in larger bundles to compose the individual muscle with its sheath of deep fascia.

ATTACHMENTS

Muscles are usually attached to bone or cartilage, sometimes to ligaments or skin, either directly by muscle fibres or by an intervening *tendon*, a sinew or 'leader', a cord-like structure of white, fibrous tissue. Most muscles have a tendon at one or both ends. Flat muscles have sheet-like tendinous expansions or *aponeuroses*, e.g. those of the flank which extend from lower ribs to pelvis. When a muscle contracts, one attachment remains fixed – the *origin* – and the other – the *insertion* – is moved towards it. Generally, in the limbs the origin is proximal and the insertion distal, while in the trunk the origin is medial and the insertion lateral. But origin and insertion may be interchangeable. Thus the pectoralis major, inserted into the humerus, normally approximates the arm to the side, but if the arm is fixed by gripping a table, the same muscle pulls on the ribs at its origin, expanding the thoracic cavity as an accessory muscle of respiration, as in asthmatic attacks. Again, the hamstrings usually bend the knee; but in straightening up from touching one's toes the knee remains straight and they extend the hip.

Attachments are usually just distal to the joint moved, e.g. the biceps into the upper radius below the elbow; this is inefficient as mechanical leverage but allows speed of action. A muscle may have two or even three origins, or *heads*, (cf. the biceps and triceps in the arm), but the insertion is nearly always single. Tendinous attachments to bone produce a ridged elevation, while direct insertions leave the bone smooth. Occasionally, muscles are attached to each other, usually between a pair on either side of the midline; thus the mylohyoids under the chin criss-cross at a line of junction, while the aponeurotic fibres of the abdominal muscles interlace in herring-bone fashion at the midline of the abdominal wall (Figures 8.6, 8.7). A few muscles have two separate bellies connected by an intermediate tendon, tethered to an adjacent bone, to redirect the pull. Some skeletal muscles act on soft tissues, e.g. those that move the eyeball and soft palate and wrinkle the skin of the face and neck.

MUSCLE FORM

The *form* of a muscle depends on the arrangement of its fibres. When these have a direct pull the muscle may be fusiform (with a tapering belly) strap-like, quadrilateral or triangular (Figure 4.8). More power is obtainable when the fibres are inserted into tendinous prolongations within the muscle substance to give a concentration of short, efficient fibres, as in the unipennate, bipennate and multipennate arrangements illustrated.

A strap-like muscle may be segmented transversely by tendinous intersections, e.g. the rectus abdominis of the abdominal wall. In the hand and foot, where bulky muscles would obstruct movement, the bellies of the muscles moving the fingers and toes are contained in the forearm or leg, and continued as slender tendons into the extremities.

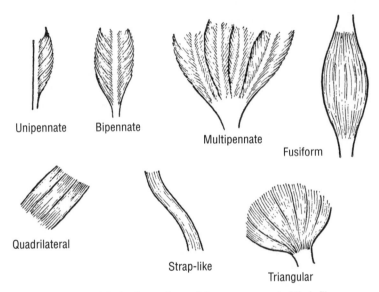

Figure 4.8 Variations in the form of skeletal muscles and the arrangement of their fibres

Muscles are named (in Latin) mainly after their actions, e.g. the *flexor pollicis longus*, the long flexor of the thumb, occasionally by shape, e.g. the *pronator quadratus*, a square muscle pronating the forearm, or by some feature, e.g. the *quadriceps* of the thigh with four heads of origin.

Muscle groups and movements

Muscles are usually grouped, with the same nerve supply, performing common or related actions It is this *movement* that is represented in the brain. No individual muscle can be contracted by an act of will, nor can any muscle act alone. We try to lift our arm, *not* to contract our deltoid, though this is the main muscle concerned, but a whole host of associated muscles come into play. In relation to a movement, a muscle may fill one of the following roles:

1 the *prime mover* or *agonist*, which is directly responsible for the movement

2 the *antagonist*, which has the reverse action and has to relax reciprocally as the prime mover contracts, paying out slack to let the latter do its job

3 the *fixators*, associated muscles steadying the base or fulcrum against which the prime mover acts

4 the *synergists*, which control an intermediate joint so that the prime mover may act efficiently; thus, the muscles closing the fingers can only give a strong grip if the synergist muscles bend the wrist backwards.

In different movements the same muscle may play different parts.

In fine movements, such as writing, both prime movers and antagonists are in some degree of contraction throughout, a balance of opposing forces; in rapid forceful movements, such as striking a blow, both groups relax once the stroke is initiated and the movement continues by momentum.

Test yourself

1 What is another name for smooth muscle?
   a Skeletal
   b Visceral
   c Striated
   d Cardiac

2 Which of the following muscles are under voluntary control?
   a Cardiac
   b Smooth
   c Skeletal
   d Plain

3 What is the thin filament in myofilaments called?
   a Myosin
   b Sarcoplasm
   c Actin
   d Mitochondria

4 What does hypertrophy of muscles mean?
   a Increase in size of muscle fibres
   b Increase in number of muscle fibres
   c Decrease in size of muscle fibres
   d Increase in excitability of muscle fibres

5 What term is used for a group of muscle cells supplied by a single neurone?
   a Electromyogram
   b Sarcolema
   c Reflex arc
   d Motor unit

6 What term is used when a muscle is unexcitable?
   a Threshold
   b Refractory period
   c Stimuli
   d Automatic contraction

7 Which part of the muscle usually moves during a muscle contraction?
   a The origin
   b The insertion
   c The ligament
   d The fascia

8 What are the ring-shaped bands of smooth muscle located at the entrance and exit of hollow organs called?
   a Sphincter
   b Tonus
   c Atrophy
   d Myosin

9 Which muscle form has a tapering belly?
  a Bipennate
  b Strap-like
  c Quadrilateral
  d Fusiform

10 What is the role of the fixator?
  a It is directly responsible for the movement
  b It has to relax reciprocally as the prime mover contracts
  c It steadies the base or fulcrum against which the prime mover acts
  d It controls an intermediate joint so that the prime mover may act efficiently

5

Standard positions, terms and references: body coverings and body systems

In this chapter you will learn about:

▶ *the anatomical terms for positional points of reference*

▶ *the structure and function of skin and fascia*

▶ *introductory information about the digestive, urinary, vascular, respiratory and lymphatic systems.*

Positions, terms and references

For purposes of anatomical description, the body is always considered to be in a conventional position: erect, with the arms by the sides and the palms facing forwards, i.e. *supinated*.

The front of the body and limbs is the *anterior* surface, the back is *posterior*, so that for any two structures, one may be described as anterior or posterior to the other, insofar as it is nearer the anterior or posterior surface. Anterior is sometimes called *ventral*, and posterior *dorsal*.

The position of structures is also related to the median plane of the body (AA in Figure 5.1). One point is *medial* to another which is further from the midline; the second point is *lateral* to the first. Points nearer the head end are *superior* to those *inferior* ones nearer the feet; *cranial* is sometimes used for superior and *caudal* for inferior.

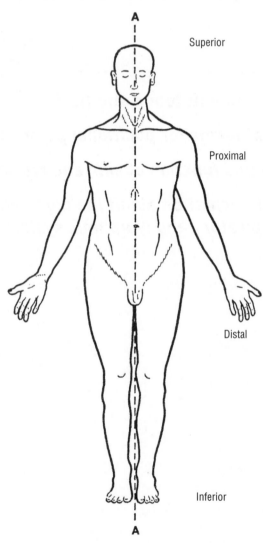

Figure 5.1 The conventional view of the body from in front, showing its anterior surface (AA is the median plane)

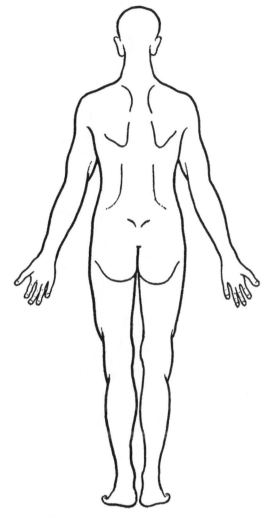

Figure 5.2 The posterior surface of the body

Internal and *external* are descriptive of the boundary walls of body cavities or hollow organs; thus the ribs have an external surface facing outwards and an internal surface towards the thoracic cavity. *Superficial* and *deep* relate to relative distances from the skin surface, e.g. the skin is superficial to the muscles.

Certain other terms are used in connection with the limbs. Here, anterior may become *palmar* (hand) or *plantar* (sole of the foot), with *dorsal* for the posterior surface, the back of the hand and upper surface of the foot. In the limbs, medial and lateral may be replaced by names derived from the paired bones of forearm and leg, i.e. *radial* and *ulnar*, or *fibular* and *tibial*. Points nearer the shoulder and groin are *proximal* to those *distal* ones situated nearer the fingers and toes. *Peripheral* roughly corresponds to distal, but is usually employed for the outlying distribution of the branches of the circulatory and nervous systems.

It is often necessary to refer to sections or planes through the body. These may be *horizontal* (transverse), *sagittal* (along or parallel to the median plane) or *coronal* (along or parallel to the

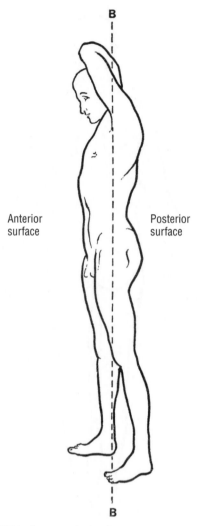

Figure 5.3 The body in side view (BB is the coronal plane)

coronal plane (BB in Figure 5.3), at right angles to the median plane). All these terms must be mastered before proceeding; they are the cardinal points of the anatomical compass.

There is an agreed system of names for the parts of the body, based on the original Latin terminology.

SYMMETRY AND SEGMENTATION

The two halves of the body on each side of the median plane are similar, with corresponding right and left limbs, right and left kidneys and so on, i.e. many structures are *symmetrical*. However, some internal organs are mainly or entirely one-sided, e.g. the liver and the spleen, while the two halves of the brain are never exactly the same.

The human body repeats, in very modified form, the primitive arrangement of exemplified *segmentation* in the earthworm, consisting of a number of identical segments, each containing the same organs. Integrated as it is, the human body retains features of this pattern, seen in the arrangement of the vertebral column, the series of paired ribs, and the segments of the spinal cord, each of which gives off a pair of spinal nerves to the appropriate body segment.

Skin

The skin (Figure 5.4) covers the body and is continuous with the linings of the internal canals at their orifices of entry and exit. It is elastic and mobile, except where bound down on the scalp, ears, palms and soles, and has a surface area of some 1.6 square metres (2 square yards). The hair follicles, sweat and sebaceous glands open on its surface and its brown pigment, melanin, is most developed on exposed sites and at the genital and nipple areas. It contains the peripheral endings of sensory nerves, acts as an excretory agent through its glands, and helps regulate body temperature through water loss by evaporation. It is also protective, and is modifed to form the hair and nails.

Microscopically, the skin has two main zones.

1   The superficial *epidermis*, thickest in the palms and soles, and permanently creased opposite the flexures of joints. Five layers can be recognized (see Figure 5.5). It has an outer *horny layer* of dead or dying flattened cells, continually shed, and renewed by growth from the *germinative layer* (see Figure 5.4). The *keratin* of the horny layer is characteristic of skin and derived structures, such as hair and nails. Within the *basal layer* a specialized group of cells produces melanin pigment. In dark-skinned individuals pigment is present in all the basal cells.

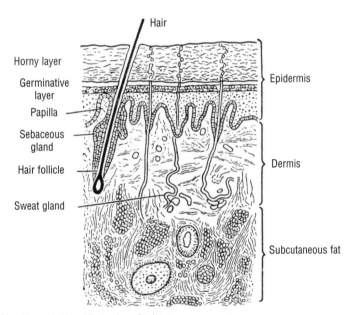

Figure 5.4 Section through the skin to show its layers

2   The deeper *dermis*, or true skin, is a layer of tough elastic and fibrous connective tissue between the epidermis and subcutaneous fat. Unlike the epidermis, it is highly vascular and projects into the epidermis in little bays, or *papillae*, which contain the terminal capillary loops and the end-bulbs of the sensory nerves for touch, pain, heat and cold. Although the hair follicles, sweat and sebaceous glands lie in the dermis, they develop as ingrowths from the epidermis, and the hair shafts and sweat gland ducts traverse the latter to reach the surface. The sebaceous glands open alongside the hairs to ensure their lubrication.

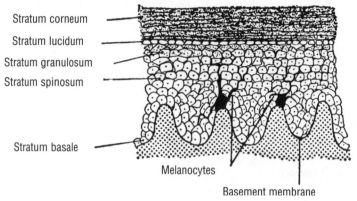

Stratum corneum

Stratum lucidum

Stratum granulosum

Stratum spinosum

Stratum basale

Melanocytes

Basement membrane

Figure 5.5 The detailed structure of epidermis (from *A Companion to Medical Studies*, Vol. 1)

The three to four million *sweat glands* occur over the whole skin surface, most numerous on the palms and soles. They are simple coiled tubular glands, well supplied with blood vessels. *Apocrine* glands are modified sweat glands in the armpit, anogenital area and breasts, related to sexual activity; they begin to function at puberty and produce the characteristic body odour. The individual *hair* consists of a long, dead, keratinized shaft and the basal growth segment or *bulb*, invaginated by the highly vascularized hair *papilla*, of specialized tissue. Strands of involuntary muscle attached to the follicles cause erection of hairs, or 'goose flesh' in cold or fear. Hair growth is intermittent. After a growth phase the lower part of the follicle degenerates, the shaft is loosened and shed as it is pushed out by the growth of a new hair.

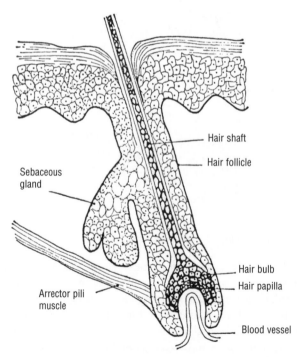

Hair shaft

Hair follicle

Sebaceous gland

Hair bulb

Hair papilla

Arrector pili muscle

Blood vessel

Figure 5.6 Diagram of hair follicle and related sebaceous gland (from *A Companion to Medical Studies*, Vol. 1)

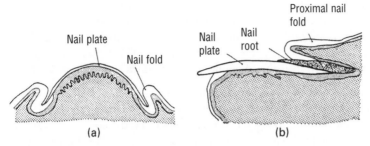

Figure 5.7 Diagram of nail: (a) transverse; (b) longitudinal (from *A Companion to Medical Studies*, Vol. 1)

The *sebaceous glands* are found everywhere except the palms and soles. They are lobulated glands whose secretion, the *sebum*, lubricates the skin. Sebaceous secretion increases at puberty under the influence of the sex hormones.

The *nail* is a translucent keratin plate embedded in folds of skin. It is free at the tip, but elsewhere it is firmly attached to the underlying epidermis. The *nail bed*, from which growth occurs, is the epidermis underlying the basal fold. Nail growth is faster in the fingers than the toes.

Nugget

Nail growth continues for a time after death.

The protective function of the skin depends on an intact epidermis. Substances may pass into the body directly via the epidermis or through the openings of the hair/sebaceous system. Penetration depends on the water and lipid solubility of the agent. The skin is almost impenetrable to water, but fat-soluble substances such as alcohol enter by dissolving the lipids of the cell walls; absorption is aided by increased skin temperature and blood supply, or if the skin is damaged. The intact skin is virtually impermeable to electrolytes and only slightly permeable to gases. In amphibia, the skin is an important respiratory organ, but in humans, cutaneous respiration constitutes only 0.5% of total (lung) respiration.

FASCIAE

Intervening between the skin and the muscles are two important layers of connective tissue, the superficial and the deep fascia.

The *superficial fascia*, the ordinary subcutaneous fat, is a continuous sheet over the whole body, everywhere except at the eyelids and male genitals. In a few places it contains muscle fibres, such as the muscles of facial expression and that which corrugates the scrotum. Fat is more developed in the abdomen, breasts and buttocks, and thicker in women. It insulates the body against cold and contains the cutaneous nerves and vessels on their way to and from the skin.

Where the superficial fascia is abundant, as in the thigh, the skin moves freely over the deep structures; where it is virtually absent, as over the nose and ear, the skin is firmly tacked down. The thicker deposit in women gives them their rounded contours; this distribution is a secondary sexual characteristic. In the male, fat tends to accumulate in later life in the abdominal wall. At certain sites – the pads of the fingers and toes, the heel-pad and the buttocks – the fat is honeycombed by fibrous strands, a cushioning device against pressure.

Internal to the superficial fascia is the *deep fascia*, a membranous sheet that covers and partitions the muscle groups and lies in close relation to bones and ligaments. From its deep surface, other sheets, or *septa*, extend inward between the muscle groups and form sheaths for nerves and vessels and compartments for the viscera. This fascia varies greatly at different sites; it is virtually absent over the face, but extremely thick in the lower back.

The general arrangement of fasciae is seen in the cross-section of a limb (Figure 7.24). This shows the skin, the superficial fascia, and the deep fascia enclosing the muscle masses in a restraining envelope and partitioning them with septa that demarcate the various groups and reach the bone to blend with the periosteum. This is important for the return of blood and lymph from the limbs. The heart pumps arterial blood into these compartments; the return of fluids towards the chest is due to the pumping action of muscular contraction within the unyielding fascial envelope. As both veins and lymphatics have one-way valves, this pushes their fluid content toward the heart, i.e. there is a varying tension within the deep fascial compartments, and this may rise after fractures to levels which endanger the local circulation – the 'compartment syndrome'. If muscular contraction is eliminated by paralysis or immobilization, fluids accumulate in the limbs, the swelling known as *oedema*. In various situations the deep fascia forms restraining bands, or *retinacula*, which hold down the tendons and prevent them from bowstringing when their muscles contract. *Synovial sheaths* facilitate the smooth gliding of tendons. *Bursae are* simple synovial sacs located over points of friction, as between skin and bony points (knee-cap and elbow) and between tendons and bones.

Body wall and body cavities

The body wall, or *parietes,* encloses the great cavities, abdomen and thorax, which are also named after their lining membranes as the *peritoneal* and *pleural* spaces. The body wall consists of the skeleton with its attached muscles and connective tissues, and the overlying skin and fat – the parietal or *somatic* structures.

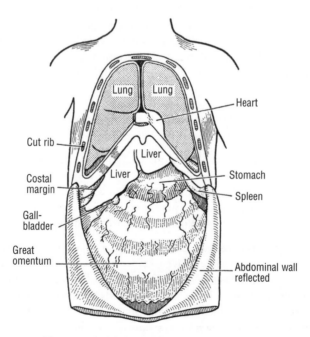

Figure 5.8 The viscera exposed by removal of the anterior walls of abdomen and thorax

The *cavities*, with their smooth serous linings, contain the internal, *visceral* or *splanchnic* organs: lungs and heart in the chest, intestines and other organs in the abdomen. The viscera develop in the embryo from the posterior body wall and retain this attachment in the adult, the lungs by their roots, the intestine by its double-layered, fat-laden, supporting *mesentery*.

The body systems

These are:

The skeletal system – the bones
The joints or articulations } the *locomotor system*
The muscles

The respiratory system
The digestive system } the *visceral organs*
The urogenital system

The vascular system – heart, blood vessels, lymphatics

The nervous system and sense organs

DIGESTIVE SYSTEM

Food passing from the mouth enters an expanded cavity behind, the *pharynx*, which is common to the airway at this level. It then travels down the gullet (*oesophagus*) to the

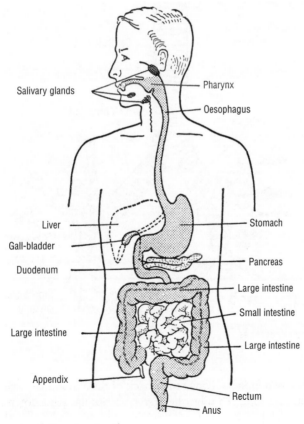

Figure 5.9 Digestive system (alimentary tract)

stomach, thence to the *small intestine* and then into the large intestine or *colon*. The wastes from the food finally reach the lowest part of the large bowel, or *rectum*, and are expelled through the short *anal canal*. At various points along the digestive tract certain glands discharge their secretions: the three pairs of *salivary glands* around the mouth, the *pancreas*, below and behind the stomach, and the *liver*, which overlies the stomach.

RESPIRATORY SYSTEM

Air inhaled through the nose or mouth enters the pharynx and travels down into the air-passages proper. The first part is the *larynx*, also the organ of voice; this leads on to the windpipe (*trachea*), which divides in the upper chest to a right and left *bronchus* for the lungs. Each bronchus subdivides in the lung into numerous branching *bronchioles*, which end in clusters of tiny *air sacs*. It is in the walls of these sacs that interchange occurs between the gases dissolved in the blood and those of the inhaled air.

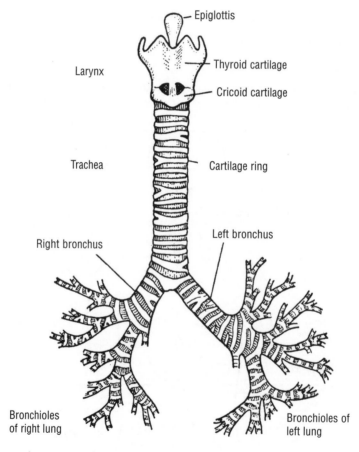

Figure 5.10 The respiratory passages

URINARY SYSTEM

Urine is secreted by the two *kidneys*, lying at the back of the abdominal cavity on the posterior abdominal wall. From each kidney a tube, the *ureter*, carries the urine down to the *bladder* in the pelvis, between the hip bones, and it is discharged to the exterior via the *urethra*. In females this is a short channel, soon opening externally; in the male it is a long pathway traversing the *prostate* gland and then the *penis*, which is also used for the reproductive act. The associated *genital* or sex organs are considered later (see Chapter 22).

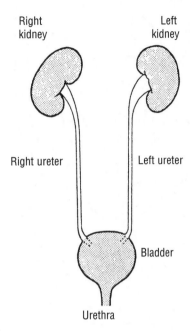

Figure 5.11 The urinary system

VASCULAR SYSTEM

The circulatory system is a closed circle, round which the blood is driven by the contractions of the heart. Blood is forced into the *arteries*, thick elastic tubes whose recoil aids its distribution peripherally. The arteries divide into smaller branches or *arterioles* on their way to the organs and limbs, and finally break up into a meshwork of fine *capillaries*, microscopic vessels that permeate every tissue of the body except the cornea of the eye, the outer layer of the skin and articular cartilage.

The blood in the capillaries discharges oxygen and food products to the tissue cells and takes up carbon dioxide and wastes in return. The network reforms to form small veins or *venules*, which become large venous trunks as they travel towards the heart. The veins are thin-walled, have no pulse, and contain valves to prevent backward flow.

There are two separate circulations: a *systemic*, concerned with the body as a whole and driven by the left side of the heart, and a *pulmonary*, concerned with passage of the blood through the lungs and driven by the right side of the heart.

The right and left sides of the heart are shut off from one another. Each has an upper chamber, or *atrium*, receiving blood from the great veins, and a valve leading to a lower chamber or *ventricle*, discharging blood into the great arteries. Stale venous blood from the body enters the right atrium, passes to the right ventricle, and is expelled through the pulmonary artery to traverse the capillaries of the lungs. Here it becomes aerated, receiving fresh oxygen in the air sacs and giving up carbon dioxide to be exhaled. The fresh blood returns from the lungs in the pulmonary veins to the left atrium, then down to the left ventricle, and is discharged into

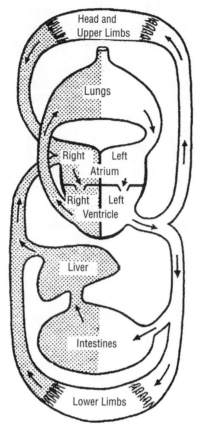

Figure 5.12 A diagram of the circulation of the blood

the great artery of the body, the *aorta*, which supplies the head, trunk and limbs through its branches.

In the tissues the blood is rendered dark and venous, and is ultimately collected into great veins – the *superior vena cava*, draining the head and arms, and the *inferior vena cava*, from the trunk and legs. Note that, whereas the arteries of the body contain bright red blood and the veins dark blood, the reverse is the case for the pulmonary arteries and veins, since the lungs reverse the chemical states of the blood.

A special arrangement of the abdominal vessels must be noted. Whereas the veins leaving most structures pass directly to the heart, those from the stomach and intestines form a *portal vein* which enters another organ, the liver, where they break up into a second set of capillaries, filtering the blood before it can reach the heart. This ensures that the liver utilizes and stores the food breakdown products brought from the bowel, and is known as the *portal circulation*. The liver capillaries reform as the hepatic veins to join the inferior vena cava.

LYMPHATIC SYSTEM

This is an accessory to the main blood vascular system. Not all the fluid part of the blood that exudes into the tissues from the capillaries returns to these vessels, so tissue fluid accumulates. This excess is removed by a separate set of fine channels, the *lymphatics*, which begin as crevices between the cells and form a plexus draining the various organs.

These vessels pass up the limbs and trunk. They have one-way valves and are relayed at certain points by *lymph nodes*, filters lying at the elbow and knee, armpit and groin, and in the trunk along the great vessels.

The lymphatics of the trunk join to form a wider vessel, the *thoracic duct*, of matchstick size, which runs up in the chest to the left side of the neck and discharges into the great veins. On the right side, the vessels of the arm, head and neck discharge directly into the veins.

One of the main functions of this system is the absorption of digested fat via the lymphatics of the bowel. The lymph nodes deal with any infection arriving via the lymphatics and also form sites for the spread (metastasis) of malignant disease.

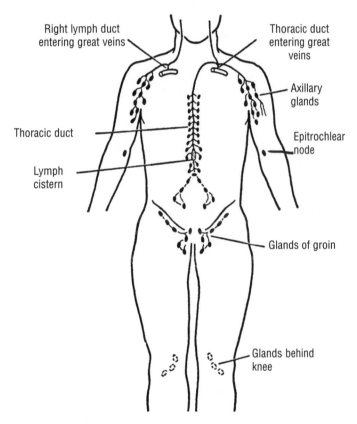

Figure 5.13 The lymphatic system (the glands of the head and neck are not shown)

Test yourself

1 What term is used for the front of the body?
   a Posterior
   b Dorsal
   c Anterior
   d Medial

2 What is the anatomical term for body points nearer the feet?
   a Lateral
   b Inferior
   c Superior
   d Cranial

3 Where are sebaceous glands *not* found?
   a Neck
   b Under arms
   c Palms
   d Knee

4 What is superficial fascia?
   a Subcutaneous fat
   b Superficial epidermis
   c Germinative layer
   d Horny layer

5 Where are visceral organs found?
   a The nervous system
   b The vascular system
   c The skeletal system
   d The digestive system

6 Which of the following tissues contain capillaries?
   a The cornea of the eye
   b The outer layer of skin
   c Articular cartilage
   d Hair

7 Where does blood come from that enters the superior vena cava?
   a The arms
   b The trunk
   c The legs
   d The feet

8 What is the correct order of blood flow between blood vessels?
   a Veins, venuoles, arteries, arterioles, capillaries
   b Arteries, veins, venuoles, arterioles, capillaries
   c Arteries, arterioles, capillaries, venuoles, veins
   d Arterioles, capillaries, arteries, veins, venuoles

9 What forms the hepatic veins?
    a Pulmonary capillaries
    b Liver capillaries
    c Alveoli capillaries
    d Coronary capillaries

10 Where are lymph nodes located?
    a Ankle
    b Wrist
    c Elbow
    d Thigh

Regional anatomy: the upper limb

In this chapter you will learn about:

▶ *the bones of the upper limb*

▶ *the joints of the upper limb*

▶ *the movements permitted at the joints in the upper limb*

▶ *the surface anatomy of the upper limb*

▶ *the muscle groups of the upper limb*

▶ *the blood vessels of the upper limb*

▶ *the nerves of the upper limb.*

In anatomy, the *arm* means the upper part of the limb between shoulder and elbow; and the *forearm* is always referred to as such.

Bones of the upper limb

SHOULDER

This consists of the rounded shoulder-cap, the prominence formed by the head of the *humerus* and the overhanging *acromion process* of the *scapula* (shoulder-blade). It also includes the scapular region behind, the pectoral region or front of the upper part of the chest below the *clavicle* (collar-bone), and the *axilla* or armpit between the two. The shoulder girdle consists of scapula and clavicle, articulating at the *acromioclavicular joint*. The medial end of the clavicle articulates with the sternum at the *sternoclavicular joint*.

The *scapula* consists of the main *body*, or blade, a thin triangular plate carrying certain projections or processes. The body has medial (vertebral), lateral (axillary) and superior borders, with a superior and inferior *angle* at either end of the medial border. But where the corresponding lateral angle would be expected, at the junction of the lateral and superior borders, there is the expanded mass of the *head* of the scapula, hollowed out on its lateral aspect to form the shallow *glenoid fossa*. The deep (anterior) surface is slightly concave and is applied to the backs of the upper ribs; the superficial (posterior) surface carries a prominent ridge, the *spine*, which runs up and out from its root on the medial border to end as a lozenge-shaped expansion, the *acromion process*, overlying the shoulder joint. Above and below the spine are the hollowed *supraspinous* and *infraspinous fossae* respectively. Lastly, there is the

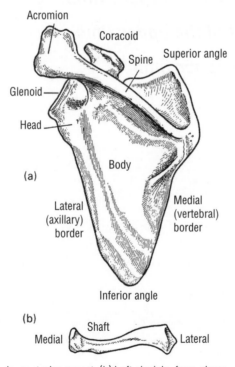

(a)

(b)

Figure 6.1 (a) Left scapula, posterior aspect; (b) Left clavicle, from above

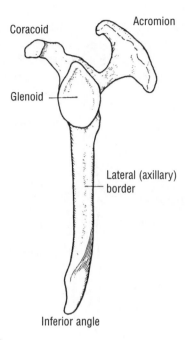

Coracoid

Acromion

Glenoid

Lateral (axillary) border

Inferior angle

Figure 6.2 Left scapula, lateral view

stubby *coracoid process*, like a crook, arising from the superior border and overhanging the joint in front. Figure 6.2 shows how the acromion and coracoid together form a protective arch or *secondary socket* over the joint.

The *clavicle* is a slender rod connecting the acromion to the upper part of the sternum, the *manubrium*. It lies horizontally, forming the lower boundary of the neck at each side. Its ends are somewhat expanded and the shaft has an S-shaped curve. Fractures of the clavicle are common, usually due to a fall on the outstretched hand, and mend rapidly.

Nugget
Total removal of the clavicle bone seems to have little effect on shoulder function.

ARM
The *humerus*, the bone of the upper arm, articulates with the scapula at the shoulder and with the forearm bones at the elbow. It has a *shaft* and two expanded *ends*. The upper or proximal end carries the rounded head, two-thirds of a sphere, directed medially and outwards. On the lateral aspect, opposite the head, are two prominences, the *greater* and *lesser tubercles*, giving attachment to the small rotator muscles which surround the joint. The greater tubercle forms the point of the shoulder beneath the overhanging acromion. The tubercles are separated by the *bicipital groove*, which carries the tendon of the long head of the biceps muscle from its origin within the joint at the upper pole of the glenoid fossa, on its passage into the arm. The true or *anatomical neck* of the humerus is the narrow strip immediately encircling the head;

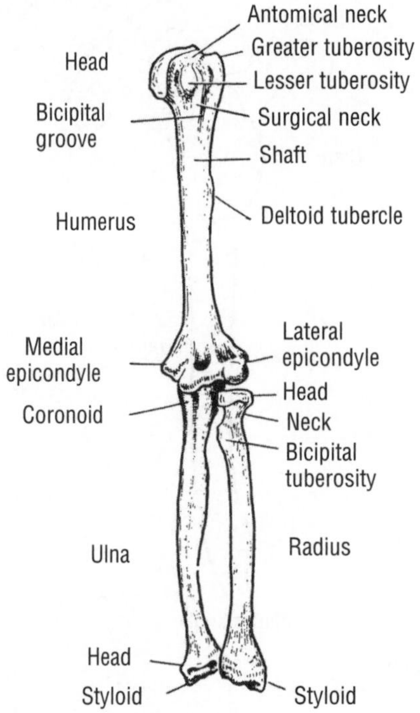
Antomical neck
Greater tuberosity
Head
Lesser tuberosity
Bicipital groove
Surgical neck
Shaft
Humerus
Deltoid tubercle
Medial epicondyle
Lateral epicondyle
Head
Coronoid
Neck
Bicipital tuberosity
Ulna
Radius
Head
Styloid
Styloid

Figure 6.3 Bones of the left upper limb, anterior aspect

the junction of the shaft proper with the whole mass of head and tubercles is the *surgical neck*, a common site for fractures in those who fall heavily on the outstretched arm.

The *shaft* is generally cylindrical but its lower third is triangular in section. Halfway down, its outer aspect shows the rough *deltoid tubercle*, the insertion of the deltoid muscle which abducts the humerus from the side. Curving round the back of the bone, just distal to the tubercle, is the *groove for the radial nerve* as it passes from the back to the front of the arm.

At the expanded lower end, the prominences above the elbow joint on each side are the *medial* and *lateral epicondyles*. The medial is more developed as it is the origin of the strong flexor muscles of wrist and fingers. The polished rounded projections from the lower end of the shaft – the lateral *capitulum* and medial *trochlea* – are concerned in the elbow joint. There are hollow depressions above them, the *radial* and *coronoid fossae* in front and the *olecranon fossa* posteriorly, which accommodate the upper ends of the bones of the forearm in extreme flexion and extension (see Figure 6.5)

FOREARM

The *radius* and *ulna* are paired bones, articulating with each other at either end to form the proximal and distal *radio-ulnar joints*. The lower end of the radius (but not of the ulna) forms the wrist joint with the carpal bones.

The *ulna* is larger and lies medially; its shaft has a sharp posterior border felt under the skin throughout the back of the forearm. The upper end carries the olecranon *process* behind, the point of the elbow, which fits into the olecranon fossa of the humerus; and the *coronoid process* in front, corresponding to the coronoid fossa. These two processes are separated by the C-shaped *trochlear notch* of the ulna which embraces the trochlear process of the

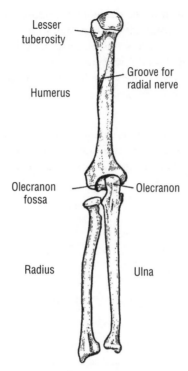

Figure 6.4 Bones of the left upper limb, posterior aspect

humerus; and just below this on the outer aspect is a shallow notch for the head of the radius. The *head* of the ulna is at its lower end, a knob on the back of the wrist carrying the small pointed *styloid process*.

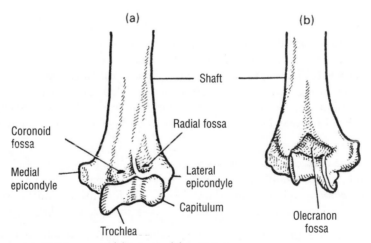

Figure 6.5 Lower end of left humerus: (a) anterior; (b) posterior

The *radius* is shorter than the ulna, lying on the lateral side of the forearm; its larger end is distal and the small *head* proximal. The head is a smooth disc, hollowed out above to receive the capitulum of the humerus, and fits within the ring formed by the radial notch of the ulna and the annular ligament of the elbow joint (Figure 6.6(a)), in which it rotates in the movements of pronation and supination (Figure 6.6(b)). The *neck* is a constriction immediately below the head and just below this on the inner side is the *bicipital tuberosity* for the insertion of the biceps. The broad lower end carries a styloid process which lies rather lower than the ulnar styloid; the back of the bone here is grooved by the extensor tendons of the wrist and fingers. The common Colles fracture occurs through the bone here, just proximal to the wrist joint.

Both forearm bones have a sharp facing *interosseous border*, connected in life by the broad sheet of *interosseous membrane* which runs the length of the forearm and separates the anterior flexor and posterior extensor compartments (Figure 6.11).

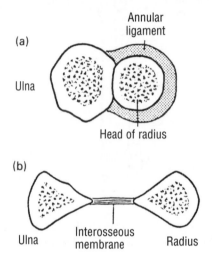

Figure 6.6 Cross-section of radius and ulna: (a) at elbow; (b) at mid forearm

WRIST AND HAND

The wrist or *carpus* consists of eight carpal bones: small irregular structures arranged in two rows of four, a proximal and a distal. These are, from medial to lateral sides: proximal row – *pisiform, triquetrum, lunate, scaphoid*; distal row – *hamate, capitate, trapezoid, trapezium*. They are bound by strong interosseous ligaments, so that while the movement at any one joint is small, the overall range gives the wrist its flexibility. The scaphoid is very often fractured by a fall on the hand and may be slow to heal. Sprains of the wrist are less common than fractures.

The skeleton of the hand consists of five *metacarpal* bones; each is a miniature long bone, with a shaft and expanded ends. The proximal *base* of each articulates with a carpal bone at a *carpometacarpal joint*, and the rounded distal *head*, which forms the knuckle, makes up the metacarpophalangeal joint with the proximal phalanx of the corresponding finger.

The fingers contain three phalanges, articulating at the two (proximal and distal) interphalangeal joints. These phalanges – proximal, intermediate and terminal – are also short long bones; each has a proximal base and a distal end formed by two condyles, making a hinge joint with the base of the phalanx in front. At the end of the finger the terminal phalanx ends in a bony tuft supporting the nail. The phalanges are concave on their anterior (palmar) surfaces because they form the floor of a tunnel roofed by fibrous tissue through which the flexor tendons of the fingers glide in their synovial sheaths.

The *thumb* is specialized. Its metacarpal is short and slight and is set freely away from the hand so that it can be opposed to the fingers. There are only two broad phalanges

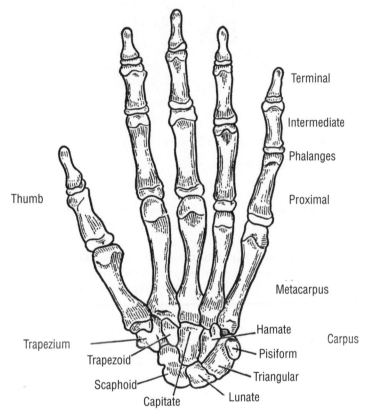

Figure 6.7 Bones of left hand, anterior aspect

Joints of the upper limb

CLAVICULAR JOINTS

The *acromioclavicular* and *sternoclavicular* joints at either end of the clavicle are simple plane joints allowing limited gliding. Each contains a projecting fibrocartilage disc, or *meniscus*. They are important mainly to complement shoulder movement. The outer joint is easily dislocated by a fall on the arm, the clavicle riding up on the acromion; the inner is very stable.

SHOULDER JOINT

Like all the upper limb joints, this is designed for mobility at the expense of stability. The large rounded humeral head does not fit well into the shallow glenoid fossa, though this socket is deepened by a fibrocartilaginous rim, the *glenoid labrum*. The joint capsule is attached round the margin of the glenoid and the head as a lax sleeve, hanging down as a fold below the joint; this laxity is necessary to allow the arm to be lifted away from the side.

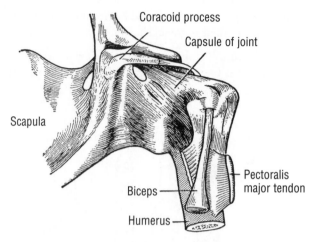

Figure 6.8 Left shoulder joint, seen from in front (note the loose capsule and the emerging tendon of the long head of the biceps)

This instability is compensated in various ways. Several thickened bands in the substance of the capsule form accessory ligaments; and the tendon of the long head of the biceps originates within the joint from the upper pole of the glenoid fossa, traversing the cavity to emerge in the bicipital groove. The main support is the blending with the capsule of the four small rotator muscles arising from the scapula as they pass to their insertion on the tubercles (Figures 6.22, 6.23, 6.24). These are the *supraspinatus*, *infraspinatus*, *teres minor* and *subscapularis*.

The muscles moving the joint also guard against displacement by tightening the capsule and holding the humeral head against the glenoid. Even so, dislocation is common, the head passing in front of the glenoid to lie under the coracoid process.

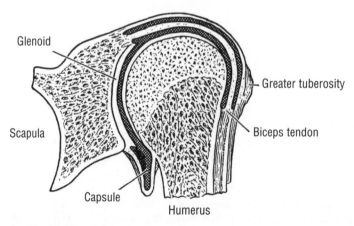

Figure 6.9 The shoulder joint, longitudinal section, anterior view (note the dependent fold of capsule below, and the biceps tendon traversing the joint cavity)

The movements of the shoulder are as follows.

1 *Abduction* and *adduction*. In *abduction*, the arm is lifted away from the side in the coronal plane by the deltoid muscle; the reverse movement of *adduction* is effected by gravity, or actively by the pectoralis major muscle in front of the chest and the latissimus dorsi and teres major muscles behind (Figure 3.10).

1 *Flexion* and *extension*, which occur in the sagittal plane, through a combination of muscles (Figure 3.10).

2 *Rotation*, produced by the small rotators arising from the scapula (Figure 3.10). It is lateral or medial, the humerus rotating on its long axis.

3 *Circumduction*, which combines the above movements in a swinging motion of the outstretched arm, which describes the shape of a cone with its apex at the shoulder.

Abduction does not occur in isolation, but in association with movement of the scapula, which can be rotated on the trunk by the large trapezius muscle of the back. The arm as a whole can be abducted through 180°, but only half of this is true shoulder motion; the rest is scapular. Both movements occur simultaneously from the outset (Figure 6.10).

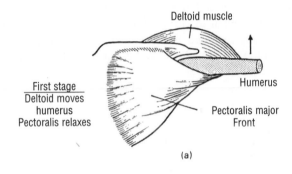

(a)

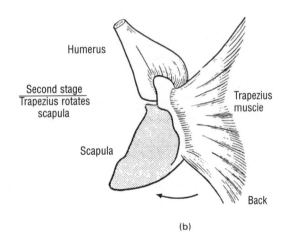

(b)

Figure 6.10 Stages in abduction of the left shoulder joint: (a) anterior; (b) posterior

These stages are not as distinct as the figures suggest, for humerus and scapula both move from the outset, the former more at the beginning and the latter more at the end of movement

ELBOW JOINT

This comprises three articulations: the true elbow hinge joint between the trochlear process of the humerus and the corresponding notch of the ulna; the shallow articulation of the radial head with the humeral capitulum; and the proximal radio-ulnar joint.

As in all hinge joints, the collateral ligaments at the sides are very strong, and with the interlocking of humerus and ulna allow only flexion and extension. The anterior and posterior parts of the capsule are lax to stretch easily at the extremes of range. When the elbow is fully extended, arm and forearm are not in line, but at an angle of 5–10°, the *carrying angle* – larger in women – which disappears as the joint is flexed. *Flexion* is produced by the biceps and brachialis muscles, the superfical and deep muscles of the anterior compartment of the arm; it is checked by soft tissue contact between arm and forearm. *Extension* usually goes a little beyond the straight position of 180° and is produced by the triceps at the back of the arm and checked by impingement of the olecranon process into its humeral fossa. The triceps tendon blends with and strengthens the back of the capsule.

The *proximal radio-ulnar joint* is formed by the round head of the radius rotating like a pivot in the embrace of the radial notch of the ulna and the annular ligament (Figures 6.6, 6.11).

Pronation and *supination* are the rotary movements of the forearm, which twist the palm backwards and forwards. The ulna remains stationary while the radius rotates round it, moving at both radio-ulnar joints and along the axis of the connecting interosseous membrane. At the inferior radio-ulnar joint the small ulnar head fits into a notch on the broad radius. Also, the bones are connected here by a triangular cartilage plate with its apex fixed to the ulnar styloid process. This apex is the lower end of the axis of rotation, the upper being the centre of the head of the radius.

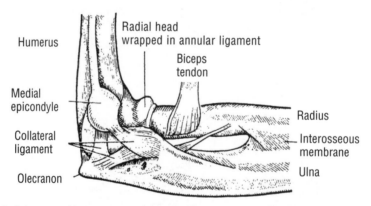

Figure 6.11 Medial aspect of left elbow joint (after *Gray*)

JOINTS OF WRIST AND HAND

The *wrist* joint proper is between the proximal row of carpal bones (scaphoid, lunate and triquetrum) and the distal end of the radius, plus its cartilage plate. It is a modified hinge joint with strong collateral ligaments and is rarely dislocated; rather, a Colles fracture is the result of a fall. Its main movements are flexion (*palmarflexion*) and extension (*dorsiflexion*). There is also movement of the wrist and hand to either side, *abduction* and *adduction* – ulnar and radial deviation – and the normal position for a strong grip is in slight adduction.

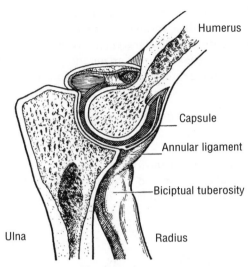

Figure 6.12 Elbow joint, longitudinal section (after *Gray*)

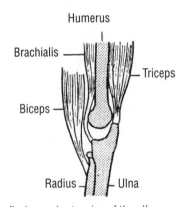

Figure 6.13 Muscles responsible for flexion and extension of the elbow

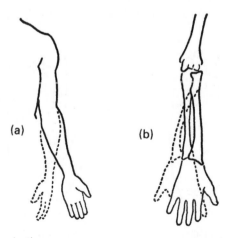

Figure 6.14 Pronation and supination

Note in (a) the 'carrying angle' in the supine position, and in (b) movement of the radius

There are numerous small carpal joints between the individual carpal bones, fastened by short interosseous ligaments, the whole forming a complicated cavity continuous with the wrist joint proximally and the *carpometacarpal joints* distally, where the metacarpals articulate with the distal row of carpal bones. Little motion occurs at the latter, except at the specialized saddle joint between the thumb metacarpal and the trapezium. Here motion is very free and consists of:

1   *adduction* to the side of the hand and *abduction* away from it, in the plane of the palm

2   movement in a plane at right angles to the palm: *palmar abduction* and *adduction*

3   the characteristic human and primate movement of *opposition*, in which the thumb is carried across the palm and its tip opposed to another digit; in this complex motion the metacarpal rotates so that the thumbnail faces forward instead of outward (Figure 6.15).

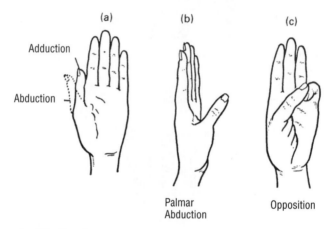

Figure 6.15 Movements of the thumb

The *metacarpophalangeal joints* between the heads of the metacarpals and the bases of the phalanges are hinge joints with strong collateral ligaments and a lax dorsal and palmar capsule. The main movements are flexion into the palm and extension backwards; usually there is 10–15° of hyperextension behind the plane of the palm. In addition there are side-to-side movements, *adduction* and *abduction*, based on an axis through the middle finger,

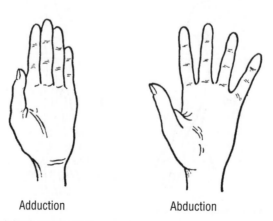

Figure 6.16 Adduction and abduction of the fingers

abduction being away from this and adduction towards it. The middle finger is abducted whether it moves laterally or medially.

The small *interphalangeal joints* allow only flexion and extension.

Surface anatomy of the upper limb

If we inspect and palpate the *anterior aspect* of the upper limb, the following features are noted.

At the *shoulder*, the *clavicle* is seen under the skin; the *acromion process* and *upper end of the humerus* give the rounded contour; and the *coracoid process* can be felt by deep pressure under the outer end of the clavicle. The bulge of the *deltoid* muscle is obvious, clothing the lateral aspect of the joint, and the *pectoral* muscles appear when the arm is adducted against resistance. In the upper arm is the bulge of the *biceps*.

At the *elbow* the prominences of the *epicondyles* are visible and palpable on either side, and the biceps belly fades into its tendon, which is felt crossing the elbow hollow, or *antecubital fossa*.

In the forearm the muscular bulge of the flexors of the wrist and fingers, enhanced by making a fist, fades into their visible or palapable tendons a few centimetres above the front of the wrist. On each side of this joint are felt the styloid processes of radius and ulna, the former slightly more proximal.

In the hand there is the palm, with a muscular eminence on either side. The larger lateral *thenar eminence*, the ball of the thumb, consists of the short intrinsic muscles of that digit, and the shallower *hypothenar eminence* medially covers the intrinsic muscles of the little finger. (Intrinsic because the digits are also moved by long tendons from the forearm.) The three main palmar creases are fairly constant, and the line of the webs bewen the fingers is not opposite the metacarpophalangeal joints, i.e. the knuckles, but at least 2 cm (2/3 inch) more distal.

Running down the *posterior aspect* of the limb, at the shoulder, the subcutaneous *spine of the scapula* ends in the overhanging *acromion process*, and here again is the bulge of the *deltoid*. The muscle mass at the back of the arm is the *triceps*, and the bony points are noted at the elbow, with the addition of the *olecranon process* of the ulna forming its point midway between the epicondyles. In the forearm, the bellies of the long extensor muscles of wrist and fingers can be seen, or felt on clenching, and at the wrist the *head of the ulna* is prominent on the dorsum. On the back of the hand the knuckles mark the *metacarpal heads* and the *extensor tendons* are seen when the forearm muscles contract.

The *surface markings* of certain underlying nerves and vessels are indicated in Figures 6.17 and 6.18. The only superficial and easily felt *nerves* are the *ulnar*, in its groove behind the medial epicondyle at the elbow, where it can be rolled to give the familiar tingling, and the *median* at the middle of the front of the wrist. Both are vulnerable to stab wounds at these sites. The main *arteries* are deep except at the wrist, where the *radial* and *ulnar* vessels can be felt pulsating on either side; it is often possible to see the beating of the superficial arterial arch in the palm. Many *veins* may be visible at the elbow and in the forearm and hand.

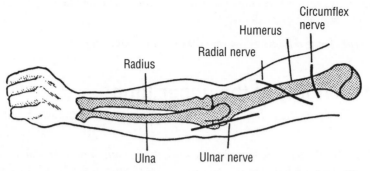

Figure 6.17 Surface markings of important vessels and nerves of the left upper limb, posterior aspect

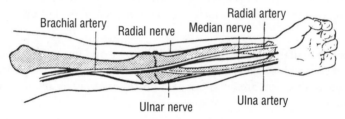

Figure 6.18 Surface markings of important vessels and nerves of the left upper limb, anterior aspect

The subcutaneous structures of the upper limb

Figure 6.19 shows the flayed limb, divested of skin and superficial fascia. Note how the deep fascia forms a continuous envelope for the deep structures. A network of veins and cutaneous nerves can be seen between skin and deep fascia, which they penetrate. The venous network in hand and forearm is gathered up into two main channels, the *cephalic vein,* runnng up the lateral side of the biceps to the shoulder, and the *basilic vein* on the medial side, piercing the fascia in front of the elbow to run with the brachial artery.

There is a *bursa* between the olecranon process and the skin, not obvious unless inflamed and distended with fluid.

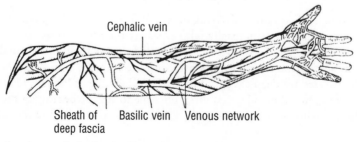

Figure 6.19 The flayed upper limb, showing the sucbutaneous structures, anterior aspect

Muscle groups of the upper limb

SHOULDER

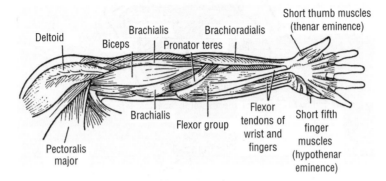

Figure 6.20 Muscles of the left upper limb, anterior aspect

The *deltoid* muscle on the outer side overlaps the front and back of the joint, in the same plane as the *pectoralis major*, which passes from clavicle and upper ribs to the upper humeral shaft. The deltoid abducts the shoulder and the pectoral adducts it. Adduction is also performed by the *teres major* and *latissimus dorsi* posteriorly, the latter being one of the great muscles of the back arising from the spinal column (see Figure 9.3). Lying very deeply, and arising from the scapular fossae, are the short rotators of the humerus, inserted into the humeral tubercles under the deltoid; they rotate the bone laterally or medially as shown in Figures 6.22, 6.23 and 6.24.

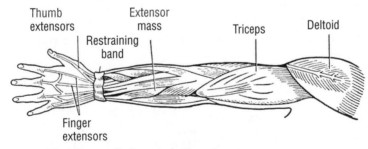

Figure 6.21 Muscles of the left upper limb, posterior aspect

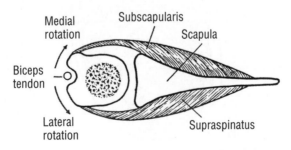

Figure 6.22 Cross-section of scapula and head of humerus to show action of short rotator muscles

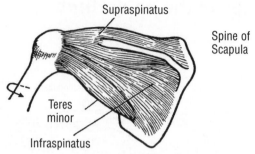

Figure 6.23 Outward rotation by the group of short rotator muscles: left shoulder, posterior view

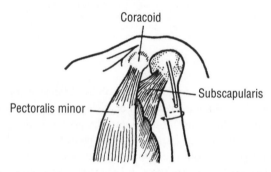

Figure 6.24 Inward rotation by the subscapularis muscle: left shoulder, anterior view

ARM

In cross-section (Figure 6.25) this is divided into *anterior* and *posterior compartments* by the humerus, wth a lateral and medial *intermuscular septum* attached to either side. The anterior (*flexor*) compartment contains muscles flexing the elbow, the *biceps* superficially and the *brachialis* near the bone.

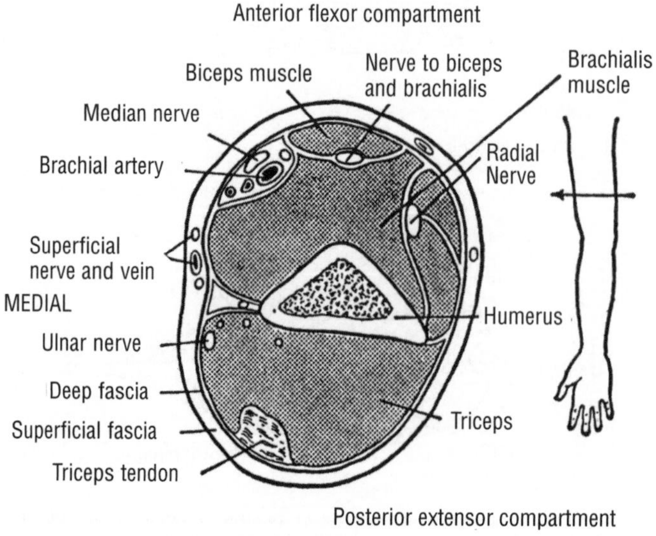

Figure 6.25 Cross-section of the upper arm near the elbow, viewed from above

The biceps has two heads of origin, a long tendon arising within the shoulder joint and a short one from the tip of the coracoid; they join to form the main belly, from which the tendon of insertion issues just above the elbow to reach the bicipital tuberosity of the radius. In addition to its flexing action, the biceps is the most powerful supinator, since it is inserted near the back of the radius and rotates it on its long axis; it is essential in resisted supination as when using a corkscrew or screwdriver. The underlying *brachialis* arises from the front of the humerus; it is an elbow flexor, inserted into the coronoid process of the ulna. In the posterior compartment is the *triceps*, which extends the elbow.

FOREARM

This too is divided into anterior flexor and posterior extensor compartments by the interroseous membrane between radius and ulna. Each compartment contains a complex group of muscles with an associated nerve trunk.

In the *flexor compartment* are a superficial and a deep group of muscles with the median nerve running between them. The *superficial group* arises from a common flexor origin on the medial humeral epicondyle. It includes the *pronator teres*, the main pronator of the forearm, inserted into the middle of the radius; the superficial flexors of the fingers; and the flexors of the wrist joint. The *deep group* arises from the forearm bones and interosseous membrane and includes the deep flexors of the fingers and long flexor of the thumb. The lateral boundary of the forearm is the *brachioradialis* (Figure 6.20), which travels from the lateral epicondyle to the lower radius, acting alternately as pronator and supinator. All these flexor bellies become tendinous 5–7 cm (2–3 inches) above the wrist, and the tendons destined for the fingers pass into the hand with the *median nerve* under a transverse restraining ligament spannng the arch of the carpal bones, an arrangement known as the *carpal tunnel* (Figure 6.27). Each finger, but not the thumb, has a superficial and a deep flexor tendon which will be studied later.

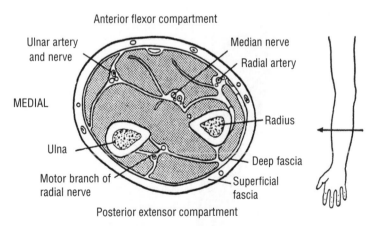

Figure 6.26 Cross-section of right forearm, viewed from above

The *extensor* forearm compartment also has superficial and deep muscle layers; and just as the superficial flexors arise from the front of the medial epicondyle, so the superficial exensors

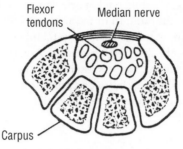

Figure 6.27 Cross-section of wrist to show the flexor tendons in the carpal tunnel

arise from the back of the lateral epicondyle. The main nerve at the back is the *radial*. The groups of extensor muscles are:

1 the extensors of the wrist joint

2 the extensors of the fingers (only one to each finger, except the index and fifth, which have two)

3 the extensors of the thumb joints.

All these tendons are bound down by a band of deep fascia or *retinaculum* at the wrist to prevent them from bowstringing backwards and to add to their mechanical leverage.

The blood vessels of the upper limb

The main artery to the upper limb begins in the base of the neck, above the clavicle, as the *subclavian*. On the right this is a branch of the brachiocephalic trunk from the aortic arch, the other branch being the right common carotid. The left subclavian and common carotid arise separately directly from the arch (see Figure 9.11).

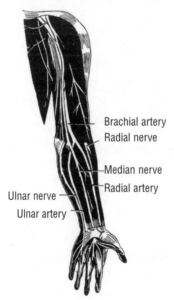

Figure 6.28 Plan of the main nerves and vessels of the upper limb

As the vessel passes out through the axilla to the arm it is called the *axillary* artery, and in the upper arm the *brachial*. Here it lies medial to the biceps and close to the median and ulnar nerves. At the elbow the brachial divides into *radial* and *ulnar* arteries, which course down the corresponding sides of the forearm in the anterior compartment. At the wrist the ulnar artery continues into the palm, while the radial turns on to the back of the carpus. In the palm there are two *palmar arterial arches* formed between branches of the two vessels, from which the *digital* vessels are given off to the fingers. A less important *dorsal arterial arch* is formed on the back of the carpus by the radial vessel alone.

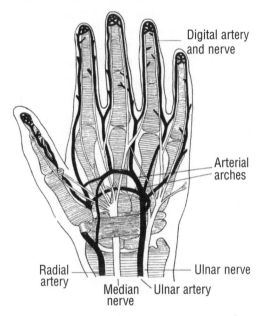

Figure 6.29 Nerves and vessels in the hand, prone (after *Bruce and Walmsley*)

The *veins* are less well-defined, forming a diffuse subcutaneous network in hand and forearm, but the main lateral (*cephalic*) and medial (*basilic*) trunks emerge at the elbow (Figure 6.19). The cephalic runs to the shoulder and dives under the clavicle to join the *subclavian* vein; the basilic vein accompanies the brachial artery to the axilla, continuing as the axillary vein and ultimately as the subclavian in the neck.

The fine *lymphatic* vessels are arranged in a superficial network corresponding to the venous pattern, and the main lymphatic trunks run with the brachial artery, entering the main group of lymph nodes of the arm in the axilla. There is a small node station halfway up the limb, in front of the medial epicondyle, the *epitrochlear* node (Figure 5.13).

The nerves of the upper limb

The nerves of the limb (Figures 6.17, 6.18, 6.28) are branches of a complex *brachial plexus* of spinal nerve roots situated in the lower part of the neck, behind the clavicle and in the axilla. From this plexus emerge the main trunks of the arm – *median*, *radial* and *ulnar* – which are grouped closely around the axillary vessels in the axilla.

The *radial* nerve runs round to the back of the upper arm, passing spirally round the humerus in its groove to emerge anteriorly just above the elbow in the fold between the biceps and brachioradialis. It then winds back round the neck of the radius to become the main nerve of the extensor compartment of the forearm. It is mainly a motor nerve, supplying the triceps, brachioradialis and extensors of wrist, thumb and fingers; but it has a small sensory supply to the skin on the back of the hand between thumb and index, and to the outer side of the forearm, and this branch runs with the radial artery in the flexor compartment.

The *median* nerve passes down the arm with the brachial artery and lies in the forearm between the superficial and deep planes of flexor muscles in the anterior compartment. It supplies most of the flexor muscles and enters the palm under the transverse carpal ligament with the flexor tendons to the fingers. In the hand it gives an important branch to the small thumb muscles, and sensory branches accompany the digital vessels, supplying the skin of the thumb, index, middle and lateral half of the third fingers on their anterior aspect. As the carpal tunnel is a small and crowded space, the median nerve may be entrapped here, giving rise to the common *carpal tunnel syndrome* of discomfort and tingling in the hand, which may require decompression by operative section of the transverse carpal ligament.

The *ulnar* nerve also runs with the brachial artery to the elbow, but enters the forearm by passing behind the medial epicondyle in the ulnar groove at the back of the humerus between the epicondyle and the trochlea, where it too may become entrapped. It runs in the flexor compartment with the ulnar artery, enters the hand superficial to the transverse carpal ligament, and supplies the small intrinsic hand muscles responsible for fine cordination of the fingers. It also gives sensation to the fifth and inner half of the ring fingers on front and back.

The hand

The main *palmar space* of the hand is bounded by the thenar and hypothenar eminences on each side; its floor is the skin of the palm, and its roof the metacarpals. Through it pass the long flexor tendons to the fingers and in it lie the arches of the ulnar artery with their branches to the fingers and the accompanying digital branches of the median and ulnar nerves. Between the metacarpals and arising from them are the *interossei*, the small intrinsic muscles of the hand. Their tendons wind round the metacarpal necks to insert into the extensor tendon expansion on the backs of the proximal phalanges; they abduct and adduct the fingers, as well as performing certain characteristic movements discussed in below.

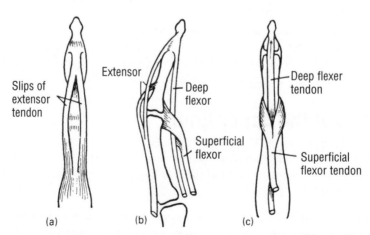

Figure 6.30 Tendon arrangements in an individual finger: (a) back of finger; (b) side view; (c) front of finger

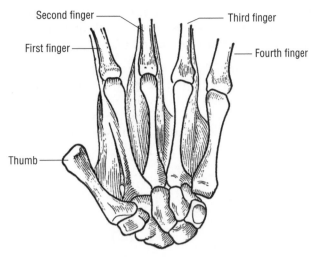

Figure 6.31 The interosseous muscles of the hand, prone

The *dorsum* of the hand is simply a shallow space between the skin and the backs of the metacarpals, traversed by the extensor tendons to the fingers.

In the *fingers* the arrangement of the flexor tendons is complex. Each digit save the thumb has two flexors, superficial and deep, which have arisen from corresponding superficial and deep muscles in the forearm. These are inserted into the phalanges as shown in Figure 6.30, the deep tendon on its way to the terminal phalanx splitting the superficial one into two halves which insert on either side of the intermediate phalanx. The tendons are facilitated in gliding by being enclosed in a smooth *synovial sheath*, and each sheath is enclosed in a fibrous tunnel, attached to the phalanges on each side (Figure 6.32). Each sheath is separate, except that of

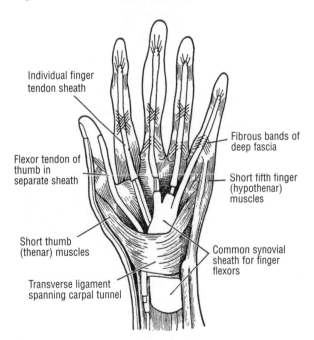

Figure 6.32 Flexor tendons and their sheaths in wrist and hand, prone (after *Gray*)

the little finger, which joins in the palm with the general common flexor sheath which envelops these tendons from just proximal to the wrist to the middle of the palm. All these tendons and the common sheath have passed through the carpal tunnel.

On the dorsum, the arrangement of the extensor tendons is simpler; there is a common synovial sheath as they pass over the back of the wrist, but the individual finger tendons are without sheaths. Figure 6.33 indicates how the tendon of each finger splits up as it is inserted. Note how the broad *extensor expansion* strengthens the capsule of the metacarpophalangeal joint, which is also the insertion of the interosseous muscles. The main features of the *nails* are shown in Figure 5.7.

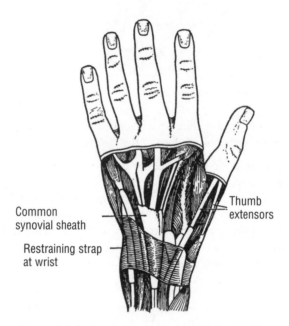

Figure 6.33 Extensor tendons and their sheaths in wrist and hand (after *Gray*)

FUNCTION OF THE HAND

In animals the digits are used purely for crude gripping and the long flexors and extensors alternately contract and relax their hold (Figure 6.34 (a and b)).

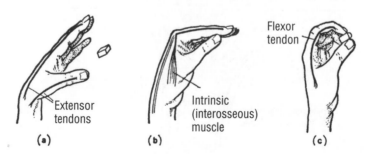

Figure 6.34 Movements of the fingers: (a) and (b) crude movements; (c) fine use of the fingers

Humans have also developed a fine coordinated action of the fingers themselves by the evolution of local intrinsic muscles confined to the hand. Their essential action for handling and working with small objects is by adoption of the writing position in the fingers and opposition of the thumb to them in a fine grip. The small muscles achieve this by flexing the metacarpophalangeal joints and then extending the interphalangeal joints by pulling on the long extensor tendons (Figure 6.34(c)).

Test yourself

1 Which two bone parts are connected by the clavicle?
   a Acromion and manubrium
   b Ribs and sternum
   c Scapula and manubrium
   d Humerus and scapula

2 Which part of the upper body carries the tendon of the long head of the biceps muscle?
   a Greater tubercle
   b Bicipital groove
   c Surgical neck
   d Lateral epicondyle

3 Identify a wrist bone on the proximal row.
   a Hamate
   b Trapezoid
   c Trapezium
   d Lunate

4 What type of joint is found at the clavicle?
   a Fixed
   b Plane
   c Cartilaginous
   d Saddle

5 Which of the following is *not* a rotator muscle of the shoulder?
   a Supraspinatus
   b Infraspinatus
   c Subscapularis
   d Glenoid labrum

6 Which movements *all* occur at the shoulder?
   a Abduction, adduction, rotation, pronation, supination
   b Abduction, adduction, flexion, extension, rotation, circumduction
   c Flexion, extension, rotation, dorsiflexion, plantarflexion
   d Abduction, adduction, hyperextension, circumduction

7 What type of joint is found at the phalanges?
   a Saddle
   b Hinge
   c Pivot
   d Ball-and-socket

8 Identify the main pronator of the forearm.
   a Brachioradialis
   b Triceps
   c Pronator teres
   d Latissimus dorsi

9 The wrist joint is located between the carpal bones and which other bone?

    a   Distal end of the ulna

    b   Proximal end of the ulna

    c   Distal end of the radius

    d   Proximal end of the radius

10 What is the name of the pulse point found at the wrist?

    a   Radial

    b   Brachial

    c   Carotid

    d   Femoral

Regional anatomy: the lower limb

In this chapter you will learn about:

▶ *the bones of the lower limb*

▶ *the joints of the lower limb*

▶ *the movements, arches, gait and mechanics of the foot*

▶ *the surface anatomy of the lower limb*

▶ *the subcutaneous structure of the lower limb*

▶ *the muscles of the thigh and leg*

▶ *the blood vessels of the lower limb*

▶ *the nerves of the lower limb.*

Anatomically, the lower limb includes the *thigh* between hip and knee, and the *leg* between knee and ankle.

Bones of the lower limb

INNOMINATE BONE

The arrangement of the pelvic girdle is shown in Figure 3.19. Each half of the pelvic ring is made up of one *innominate* or hip bone. The two bones articulate in the midline anteriorly at the *pubic symphysis*, the bony prominence felt just above the external genitalia. Behind, each bone articulates with the sacrum of the spine at the *sacroiliac joint*. Both symphysis and sacroiliac joints are strong junctions allowing little or no motion, though they relax somewhat in late pregnancy to promote delivery.

The innominate bone is a support for the lower limb, an attachment for the proximal limb muscles, and, with its partner, a bony container for the pelvic organs (see Chapter 8). It consists of three main portions – *ilium, ischium* and *pubis* – all centred on the *acetabulum*, the socket for the femoral head.

The *ilium* is a flaring broad sheet of bone whose upper border is the *iliac crest*, felt with the hand on the hip. The crest has a *spine* at either end.

The *pubis* has a superior and inferior strut or *ramus*; the two rami, with the ischium, enclose a large gap, the *obturator foramen*. On the superior ramus is a little knob, the *pubic tubercle*, which can be felt at the inner end of the groin.

The *ischium* is the dependent portion of the pelvis, carrying the broad *ischial tuberosity* which takes weight in sitting.

The *acetabular socket* of the hip joint is a deep cavity in the centre of the outer aspect of the bone, with an encircling rim shaped like a horseshoe, deficient at the *acetabular notch* in front and below.

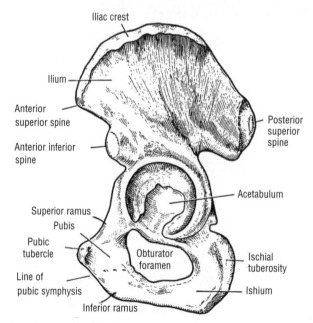

Figure 7.1 Left innominate bone, outer aspect

There are important ligaments and muscles are attached to the innominate bone. The anterior spine of the ilium and pubic tubercle are spanned by the *inguinal ligament* of the groin (Figure 7.20), separating thigh from abdomen, under which the great vessels and nerves pass into the limb. The obturator foramen is bridged by the *obturator membrane*, pierced by nerve twigs and small vessels, and the acetabular notch is bridged by the *transverse ligament*. The broad outer surface of the ilium is the origin of the gluteal muscles of the buttock; the ischium of the hamstrings of the back of the thigh; and the pubis of the adductor muscles which pull the limb towards the midline.

FEMUR

This has a long cylindrical *shaft* with slight forward and outward bowing. At its upper end the *head*, some two-thirds of a sphere, is set well off the shaft at an angle of some 130° by the long stout *neck*. At the base of the neck are the two *trochanters*, the great trochanter laterally and the small trochanter medially. The neck acts as a lever for the muscles attached at its base, so that the restricted movement which is the price of stability at the hip is compensated by an increased mechanical advantage.

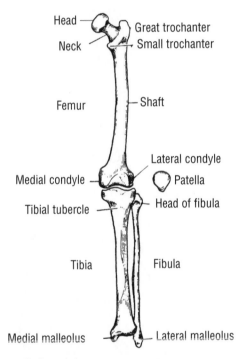

Figure 7.2 Bones of the left lower limb, anterior aspect

The shaft is thickly clothed with muscles, many attached to the rough bony ridge running down the length of its posterior aspect – the *linea aspera*. Not far below the neck the back of the bone is marked by the *gluteal tuberosity*, the insertion of the great gluteus maximus muscle of the buttock. A few centimetres above the knee the shaft expands in triangular fashion, ending in two massive *condyles*, lateral and medial, separated by an *intercondylar notch,* much deeper behind. The medial condyle is more prominent and at a slight angle to the line of the shaft as seen from below (Figure 7.4). The lower end of the femur has a smooth surface above the condyles in front articulating with the patella, and a broad posterior surface facing into the *popliteal fossa*, the hollow at the back of the knee. There is a small *epicondyle* on the outer surface of each condyle, and the *adductor tubercle* at the summit of the medial condyle is the lowest attachment of the adductor muscles.

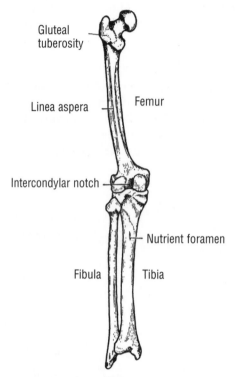

Gluteal tuberosity

Linea aspera

Femur

Intercondylar notch

Nutrient foramen

Fibula

Tibia

Figure 7.3 Bones of the left lower limb, posterior aspect

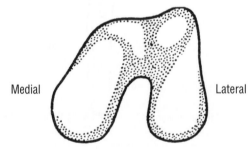

Medial

Lateral

Figure 7.4 Condyles of left femur, end-on view

The *patella* or knee-cap is a *sesamoid* bone (a sesamoid bone is situated within a tendon), in the quadriceps muscle in front of the thigh which extends the knee. It is roughly triangular, apex downwards, and has an upper and lower pole and anterior and posterior surfaces. The posterior surface articulates with the smooth femoral surface above the condyles. The quadriceps tendon is inserted into the upper pole and its pull is transmitted to the *patellar ligament*, which connects the lower pole to the tubercle of the tibia.

TIBIA AND FIBULA

These are the paired bones of the leg, on the medial and lateral sides respectively. They articulate together at the proximal and distal *tibiofibular joints*. The tibia articulates with the femur at the knee and with the talus bone of the tarsus at the ankle. The fibula participates in the ankle but is excluded from the knee joint. Each bone has an oppposing sharp *interosseous border*, connected by the *interosseous membrane*.

The *tibia* has a long shaft, triangular in cross-section, with medial, lateral and posterior surfaces, and medial, lateral and anterior borders. The anterior border is the sharp crest felt as the shin. The flat medial surface of the tibia is beneath the skin at the inside of the leg. The upper end of the bone expands as two broad flat masses, the *medial* and *lateral condyles*, which articulate with those of the femur, the semilunar cartilages intervening (Figure 7.10). The smooth condylar surfaces are separated by a rough intercondylar area carrying a projecting *tibial spine*. At the upper end of the tibial crest, below the condyles, is the prominent *tibial tuberosity*, the insertion of the extensor muscles of the knee. The lower end is much narrower than the upper and has a projecting process, the *medial malleolus*, overhanging the inner side of the ankle.

The slender *fibula* transmits little body weight. Its shaft is polygonal on section, with numerous ridges for muscle attachments. The head of the fibula is located at the upper end and has at its apex a pointed *styloid process*, the insertion of the biceps femoris muscle of the thigh and of the lateral ligament of the knee. The lower end of the bone projects as the *lateral malleolus*, overhanging the outer side of the ankle.

FOOT

The bones of the foot fall into three groups: the *tarsals*, *metatarsals* and *phalanges* of the toes. The *talus* lies immediately beneath the long bones, articulating with them at the ankle, and rests on the *calcaneus* or heel bone. But it inclines so obliquely to the inner side of the foot that the anterior ends of talus and calcaneus come to lie side by side, forming the proximal

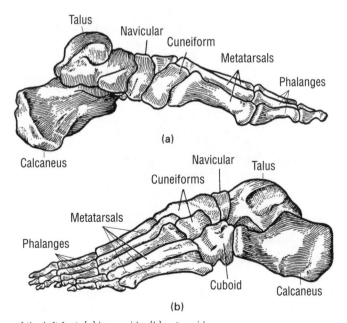

Figure 7.5 The bones of the left foot: (a) inner side; (b) outer side

row of the tarsus. The distal row consists of three *cuneiforms*, at the inner side and the *cuboid* laterally; between the two rows, on the inner side of the foot, is the *navicular* separating talus from cuneiforms.

The *talus* has a body with a rounded superior surface at the tibiofibular articulation, and a *neck* carrying the *head*, which fits into the navicular. The *calcaneus* has a projecting posterior *tuberosity*, the prominence of the heel, set at an angle to the body of the bone. It articulates in front with the back of the cuboid. Talus and calcaneus are connected by a strong *interosseous ligament*, which bridges the *subtalar joint* between the two bones.

The *cuboid* and *cuneiforms* are irregular bony masses articulating with the tarsus behind and the *metatarsals* in front. The latter resemble the metacarpals of the hand, except that the first metatarsal lies closely parallel to the others and entirely lacks the mobility of the thumb metacarpal, since the human great toe cannot be opposed as in some primates. It is stouter than the others and its head is supported underneath by a pair of tiny *sesamoid* bones. The long axis of the foot as regards abduction and adduction of the toes is the second metatarsal, not the middle bone, as in the hand. The toes have three phalanges, except the great toe with only two.

Joints of the lower limb

HIP JOINT

The deeply embedded head of the femur forms the hip joint by articulating with the acetabulum. The socket is deepened by a fibrocartilaginous rim or *labrum* round its periphery, and its deep central portion is out of contact with the head. This joint has no articular cartilage and is filled by a fatty pad. In the centre of the femoral head is a small depression, the *fovea centralis*, opposed to this non-articulating region of the acetabulum. From the fovea a stout round cord, the *ligamentum teres*, runs to the margins of the acetabular notch, carrying blood vessels to the head of the femur.

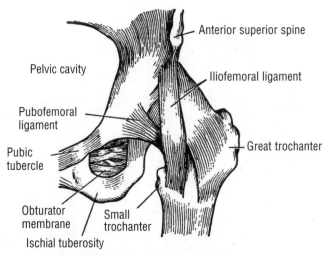

Figure 7.6 Left hip joint, anterior aspect

The capsule of the hip is strong and thick, a blending of bands from each portion of the innominate bone – *iliofemoral*, *pubofemoral* and *ischiofemoral* ligaments. It is attached proximally around the acetabular margin and reaches distally to the base of the femoral neck in the trochanteric region, i.e. the whole of the neck is inside the capsule. The synovial membrane lining the capsule is reflected upwards at the base of the neck, clothing it as far as the head.

Movements at the hip correspond to those at the shoulder. They are:

1 *abduction* and *adduction* in the coronal plane away from and towards the midline

2 *flexion* and *extension* in the sagittal plane, in front of and behind the trunk

3 *lateral* and *medial rotation*, a twisting movement of the thigh on its long axis.

The joint is liberally supplied with nerves, some of which also serve the knee, so that pain from hip disease is often referred to the knee and thought to arise from that joint.

KNEE JOINT

This is a complex joint with several communicating departments. The fibula plays no part in it. The main parts are:

1 the *tibiofemoral* joint between the tibial and femoral condyles of each side, a *menis cus* intervening on each side

2 the *patellofemoral* articulation between the back of the patella and the front of the femur.

The knee is essentially a hinge joint, with *flexion* and *extension* as its main movements, the strong collateral ligaments resisting any sideways strain. But there is an element of *rotation*, a screwing motion that locks the joint as it is finally straightened, so that in the fully extended position in standing it is very stable. There may be a small degree of *hyperextension* beyond 180°.

The loose and extensive anterior portion of the capsule is strengthened by the expansion of the quadriceps tendon as it passes via the patella to the tibial tubercle. There is a blending of tendon and capsule, so that the tonus of the muscle guards the joint against strain, keeps the capsule taut and prevents fluid from distending the capsule.

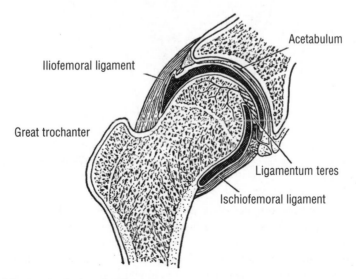

Figure 7.7 Hip joint, longitudinal section (after *Gray*)

The collateral ligaments and posterior capsule are fairly taut, but the anterior capsule is loose, especially where it extends around and above the patella as the *suprapatellar recess*, a pouch which is necessarily baggy to allow flexion of the joint. When the knee is fully extended, the patella articulates with the femur above the condyles; in increasing flexion it lies against portions of both condyles.

The synovial cavity of the knee includes:

1 the *patellofemoral space*, with the suprapatellar recess in the highest part of the anterior compartment

2 the main *anterior joint cavity* between tibial and femoral condyles on each side

3 the *intercondylar portion*, the tunnel between the femoral condyles, traversed by the two crossed *cruciate ligaments* which tie each femoral condyle to the opposite tibial condyle (Figure 7.8)

4 the *postcondylar space* on each side, a loose synovial pouch behind each femoral condyle.

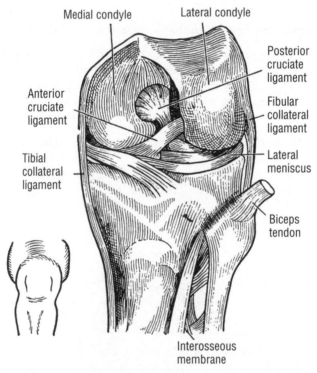

Figure 7.8 Left knee joint, anterior aspect; the joint is bent at a right angle, and the front of the capsule has been removed (after *Gray*)

The *menisci* are concentric incomplete discs intervening between the tibial and femoral condyles and fixed to the deep aspect of the capsule and the outer rim of the tibial condyles (Figure 7.10). They project for 1–2 cm (1/2–3/4 inch) into the joint, with a thin free edge. Their front and back ends are the *anterior* and *posterior horns*; those of the more elliptical medial meniscus embrace those of the circular lateral meniscus. The two menisci cushion bony contact and are bound down to the upper surfaces of the tibia, with which they rotate.

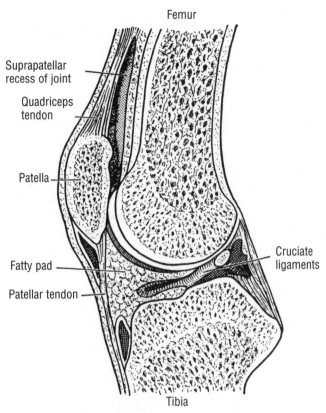

Figure 7.9 The knee joint, longitudinal section (after *Gray*)

However, they also have an attachment to the femur via the collateral ligaments, and this double fixation is responsible for the rotary strain which tears a meniscus, the common football injury, usually a longitudinal split or 'bucket-handle' tear. The inner fragment slips towards the centre of the joint and locks it by blocking full femorotibial extension. The knee is well suited to inspection by endoscopy and loose fragments can be removed by this means.

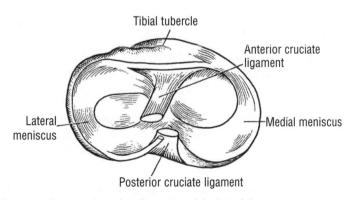

Figure 7.10 Semilunar cartilages and cruciate ligaments of the knee joint
The view is of the upper surface of the left tibia, the femur having been removed

The *tibiofibular joints* include:

1 the *superior tibiofibular articulation*, a simple plane joint formed by the fibular head abutting against the lateral tibial condyle

2 the connection of the shafts by the *interosseous membrane*

3 the *inferior tibofibular joint*, a firm fibrous unyielding union.

The fibula hardly moves on the tibia and is bypassed in the transmission of body weight.

ANKLE JOINT

This is between the upper surface of the talus and the lower ends of tibia and fibula, but weight transmission is only from the tibia. The malleoli, overhanging the talus at either side, form a *mortice* in which that bone is wedged; there are strong collateral ligaments and the only movements are those of downward *plantarflexion* and upward *dorsiflexion* of the foot. Forced lateral or rotary movements produce the common *Pott's fracture* of one or both malleoli with sideways shift of the talus. The *tarsal joints* form a complex system best considered in terms of the movements that take place.

MOVEMENTS OF THE FOOT

The foot as a whole may be *inverted* or *everted*, i.e. the sole turned inwards or outwards, and this is best understood by considering it as moving *en masse* at the subtaloid joint (the joint between talus and calcaneus and at the talonavicular ball-and-socket, with the talus acting as a stationary pivot (Figure 7.11)).

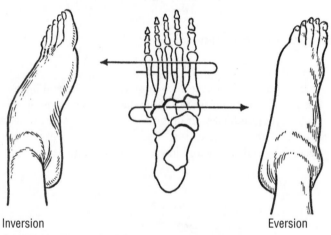

Inversion   Eversion

Figure 7.11 Inversion and eversion of the foot

The arrows merely indicate the direction of motion; both movements occur at the same system of joints

In addition, the forefoot – metatarsals and toes – may be *abducted* or *adducted*, i.e. deviated medially or laterally, keeping the sole parallel to the ground; this occurs at the *midtarsal joint*, which traverses the foot and comprises the talonavicular and calcaneocuboid articulations (Figure 7.12).

It is not really possible to separate these movements; inversion is always accompanied by some adduction, and eversion by some abduction. It is difficult to grasp that none of these sideways

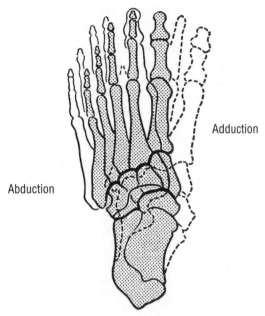

Adduction

Abduction

Figure 7.12 Abduction and adduction of the forefoot, viewed from above
The shaded position is neutral, the outlines indicate the extremes of range

and rotary movements is occurring at the ankle, which is capable only of up and down movement.

The remaining *tarsal joints* are small plane articulations and the joints of the toes are like those of the fingers.

THE ARCHES OF THE FOOT

These are longitudinal and transverse. The *longitudinal arches* lie in the long axis of the foot, with a higher *medial* and a lower *lateral* long arch, with a common posterior pillar in the calcaneus, from which they diverge. The line of the medial arch is calcaneus-talus-navicular-cuneiforms-inner three metatarsals. The line of the outer is calcaneus-cuboid-outer two metatarsals, and this is much flatter. The talus lies at the summit of the long arches, so that the body weight constantly tends to flatten them – and would do were it not for the supporting ligaments and muscles.

The *transverse arch* is the side-to-side concavity seen in cross-section and is most marked at the bases of the metatarsals. The long arches are not present at birth but develop in the first eighteen months; the transverse arch is present in the foetus. The importance of the arches in foot function is not stressed as much as it used to be. 'Flat foot' is of little importance; what matters is that the foot be supple and capable of assuming the arched position voluntarily, as in children and ballet dancers, who have excellent function though their feet may be very flat. It is rigidity or acute foot strain that is painful and disabling. An unduly high arch or hollow foot, *pes cavus*, is rigid and more troublesome.

The supports of the arches are the binding ligaments and the tone of the muscles of the calf and sole; the latter is more important. To some extent the shape of the bones contributes to their maintenance. The important ligaments are those on the underside of the foot, short bonds between individual bones or longer structures running from one pillar to the other.

The tendon of the *tibialis anterior* muscle (Figure 7.15) runs down from the leg to pull on the medial cuneiform and thus maintain the summit of the long arch. The tendon of the *tibialis posterior* of the calf curves behind the ankle and sustains the head of the talus from below as it passses to its insertion into the navicular. And the *peroneus longus* tendon from the outer side of the leg (Figure 7.18) crosses under the sole from lateral to medial side, bracing the transverse arch.

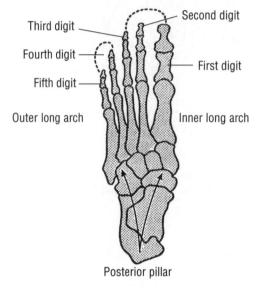

Figure 7.13 Longitudinal arches of the foot, viewed from above

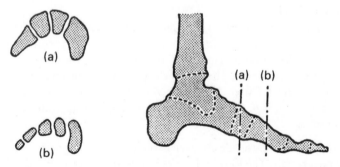

Figure 7.14 Cross-section of the transverse arch of the foot, medial view: (a) in the tarsal; (b) in the metatarsal region

Finally, the intrinsic muscular mass of the sole, the *short plantar muscles*, attached behind to the underside of the calcaneus and extending to the toes, supports both long arches.

The ligaments are not intended to take the entire strain of body weight, except momentarily; they are guarded by the calf muscles whose tendons have been mentioned above. If this tonus is weakened, the ligaments are overstretched and acute foot strain results.

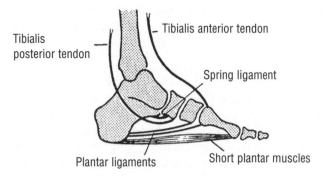

Figure 7.15 Ligaments and tendons supporting the longitudinal arches of the foot, medial view

GAIT AND THE MECHANICS OF THE FOOT

The arches are spring-like structures, yielding under body weight and with elastic recoil. The principal ligament, the short band between calcaneus and navicular which supports the head of the talus, is known as the *spring ligament*. In walking, the weight is planted on the heel, then transmitted along the outer border of the foot, and finally across the transverse arch to the first metatarsal head; and flexion of the metatarsophalangeal joints gives a kick-off onto the other foot. The remarks about primitive and fine movements of the fingers (Chapter 6) also apply to the toes. In walking on rough ground, particularly with bare feet, the long flexors act as powerful gripping agents. But on pavements and in shoes it is necessary to bring the toes flat down for efficient thrust and this is effected by the intrinsic muscles. Should this control be lost, the unopposed long flexors and extensors produce clawing of the toes, which hyperextend at the metatarsophalangeal joints and flex at the interphalangeal joints. This exposes the metatarsal heads in the sole and callosities and clawing develop, which may become permanent (Figure 7.16).

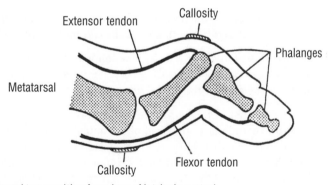

Figure 7.16 The clawed toe resulting from loss of intrinsic control

Surface anatomy of the lower limb

ANTERIOR ASPECT

The *groin* marks the transition between abdominal wall and thigh; pressure in this fold reveals the resistance of the underlying *inguinal ligament*, which can be traced laterally to the *anterior superior spine* of the ilium and medially to the small knobby *pubic tubercle*. The latter is often obscured in men by the *spermatic cord* running to the testicle.

Immediately distal to the ligament is the *inguinal region,* which the main vessels and nerves of the limb have entered from the abdomen. The pulsation of the femoral artery is felt just below the midpoint of the ligament, the femoral nerve more laterally. There are numerous lymph nodes.

In the *thigh*, the main anterior muscle bulge is fomed by the *quadriceps femoris*, tapering towards its insertion into the patella. The mass on the medial side is the *adductor group* of muscles, whose tendons of origin can be felt running up to the pubic bone, just lateral to the external genital organs. When the knee is braced straight, a firm resistance is felt under the skin on the outer aspect of the thigh from hip to knee; this is the *iliotibial band* of deep fascia, which is kept taut by the buttock muscles and helps maintain erect posture.

At the front of the *knee* is the *patella*, and in line with it about 5 cm (2 inches) below is the *tibial tuberosity*; the *patellar tendon* connecting the two and transmitting quadriceps pull becomes obvious when the knee is extended, and a hollow on either side of the tendon indicates the anterior joint compartment immediately under the skin. The femoral and tibial *condyles* are easily seen and felt at the *head of the fibula* on the outer aspect. Just below the head the *peroneal nerve* can be rolled over the neck of the fibula.

The *tibial crest*, its anterior border or shin, is subcutaneous from tibial tuberosity to ankle, with the flat subcutaneous medial surface on its inner side. On the outer side of the crest is the bulge of the muscles in the anterior compartment of the leg, and more laterally still the *peroneal muscles* obscure the underlying fibula, which emerges under the skin just above the ankle.

At the ankle the two *malleoli* stand out on either side, and 2 cm (3/4 inch) below the medial can be felt the pulsation of the *posterior tibial artery.*

The *navicular* often makes a prominence on the medial border of the foot, as does the base of the fifth metatarsal on the outer side.

POSTERIOR ASPECT

The great muscle masses at the back of the limb are, from above downward: the *gluteal* muscles of the buttock, the *hamstrings* in the thigh and the *calf* muscles of the leg.

The *buttock* has a rounded *gluteal fold* below at its junction with the thigh; following this laterally, the resistant bony mass of the *great trochanter* is felt under the skin. Deep in the lowest part of the buttock is the apex of the *ischial tuberosity*.

At the back of the *knee* the tendons of the hamstrings diverge; they can be seen or felt as the *biceps femoris* passes down and out to the head of the fibula, the *semimembranosus* and *semitendinosus* down and in to the tibia. These tendons mark the upper boundaries of the diamond-shaped space behind the knee, the *popliteal fossa*, in which lie the great vessels which have travelled round from the front of the thigh in its lower third, as well as the *sciatic nerve* and its branches.

The bulk of the calf is made up of the the gastrocnemius and soleus muscles, which taper to form the Achilles tendon, passing behind the ankle to the tuberosity of the calcaneus.

The *femoral artery* and *vein* spiral round the inside of the femur to reach the popliteal fossa. The *sciatic nerve* runs straight down the middle of the back of the thigh to divide a few centimetres above the knee, its *common peroneal* division winding round the neck of the fibula to the front of the leg and its *tibial* branch continuing down into the calf.

Subcutaneous structures of the limb

Figure 7.17 shows the flayed limb, with the superficial structures between skin and deep fascia. The venous network is gathered up into two main channels. The *great saphenous vein* runs up in front of the medial malleolus, along the medial border of the leg and thigh, and finally to the inguinal region, where it passes through an oval window in the deep fascia to join the main femoral vein deeply. The *small saphenous vein* is formed on the lateral side of the foot, passes behind the lateral malleolus up the midline of the back of the calf to the knee, where it pierces the deep fascia to join the *popliteal vein* on its way to become the femoral vein.

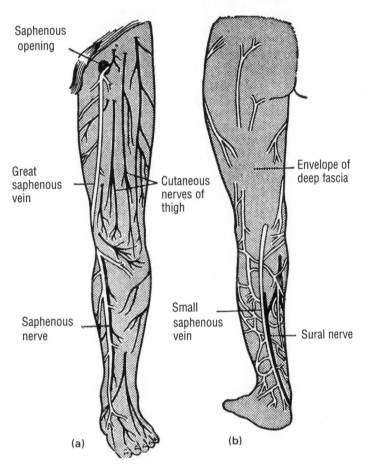

Figure 7.17 The flayed lower limb: (a) anterior; (b) posterior

Both veins are accompanied by lymphatic vessels; there are a few small popliteal *lymph nodes* at the termination of the small saphenous vein and numerous large inguinal nodes around the upper part of the great saphenous trunk at the groin.

The cutaneous nerves are many, the front of the thigh is supplied by *lateral*, *intermediate* and *medial* branches from the femoral nerve; the back by a small branch of the main *sciatic* trunk. The *saphenous* branch of the femoral nerve runs the whole length of the limb with the great

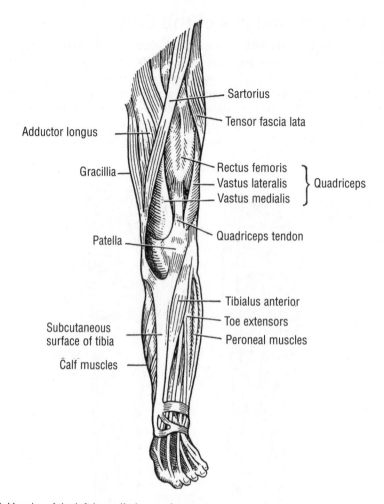

Figure 7.18 Muscles of the left lower limb, anterior aspect

Labels in figure:
- Sartorius
- Tensor fascia lata
- Adductor longus
- Rectus femoris
- Vastus lateralis
- Vastus medialis
- Quadriceps
- Gracillia
- Quadriceps tendon
- Patella
- Tibialus anterior
- Toe extensors
- Peroneal muscles
- Subcutaneous surface of tibia
- Calf muscles

saphenous vein, and the *sural* branch of the sciatic accompanies the small saphenous vein to the outer side of the foot.

The main subcutaneous *bursae* in the limb are:

1 a bursa between skin and ischial tuberosity to relieve pressure in sitting

2 a bursa over the outer aspect of the great trochanter, which may be inflamed by excessive cycling

3 the *prepatellar bursa* in front of the patellar tendon and lower half of the patella, which is not part of the knee joint and takes weight in kneeling, and its inflammation and distension is the cause of 'housemaid's knee'

4 a bursa between the skin and Achilles tendon, where a shoe rubs against the back of the heel.

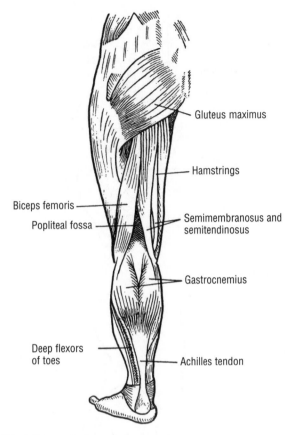

Figure 7.19 Muscles of the left lower limb, posterior aspect

The labels in the figure read:
- Gluteus maximus
- Hamstrings
- Biceps femoris
- Popliteal fossa
- Semimembranosus and semitendinosus
- Gastrocnemius
- Deep flexors of toes
- Achilles tendon

Muscles of the thigh

These fall into several main groups.

1 *Muscles connecting trunk and femur: iliacus* and *psoas* (Figure 7.20). The *iliacus* is a broad sheet arising from the inner (pelvic) surface of the iliac portion of the inominate bone. The *psoas* is a long belly at the back of the abdominal cavity alongside the lumbar vertebrae. The two muscles form a conjoined *iliopsoas tendon*, which enters the thigh by passing under the inguinal ligament lateral to the femoral vessels to its insertion at the lesser trochanter.

2 *Muscles connecting pelvis and femur.* There are two main groups: the gluteals and the adductors.

The *gluteal muscles* of the buttock arise from the outer surface of the ilium and the back of the sacrum. The most superficial, covering all the others, is the great *gluteus maximus*, which inserts into the gluteal tuberosity of the femur and the iliotibial band of deep fascia. It is an important postural muscle, for it extends the hip, carries the limb back in walking, and braces the limb by tightening the deep fascia (Figure 7.21).

Beneath it are the smaller *gluteus medius* and *minimus*, which are the main *abductors* of the hip from the midline. These are also posturally important, as they make it possible to

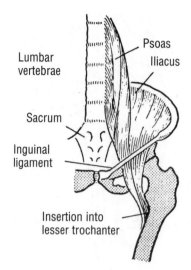

Figure 7.20 Iliopsoas group of muscles, left side

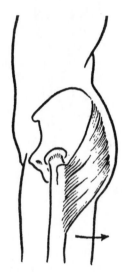

Figure 7.21 Action of gluteus maximus, lateral view

stand on one leg by tilting the pelvis over in line with the weight-bearing limb, resisting the tendency of the trunk to fall to the other side. Thus they make walking possible, for this is no more than an alternate standing on either leg (Figure 7.22).

The *adductors – longus, brevis* and *magnus* – lie on the inner side of the thigh. They arise by tendons from the pubis and ischium, running down and outward to be inserted into the shaft of the femur as far down as the adductor tubercle. They adduct the femur to or beyond the midline. The three muscles are arranged in layers from front to back in the order named, and are supplied by the *obturator nerve*, which enters the thigh from the pelvis by piercing the obturator membrane. One superficial straplike member of this group, the *gracilis*, reaches as far as the tibia.

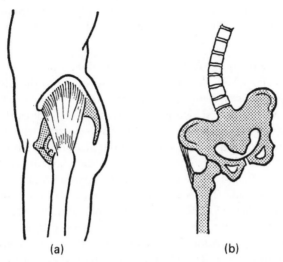

Figure 7.22 (a) The gluteus medius, lateral view; (b) the stabilizing action of the gluteus medius in standing on one leg, anterior view

3 The *short rotators* of the hip are situated deep in the buttock, close to the sciatic nerve, and are seen only after removing the overlying gluteal muscles. They arise from the sacrum, insert into the base of the femoral neck and laterally rotate the hip.

4 The *quadriceps femoris* is the great mass in the front of the thigh responsible for extending the knee. Three heads arise from the femoral shaft – the *vastus lateralis, intermedius* and *medialis* – and a fourth is the *rectus femoris* from the ilium just above the acetabulum. There is a common *quadriceps tendon* emerging in the lower thigh and inserted into the upper pole of the patella. Its expansions strengthen the capsule of the knee joint. All the heads are supplied by the femoral nerve.

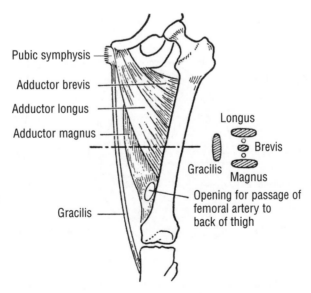

Figure 7.23 Adductor muscles of the left thigh, anterior view; their disposition is also shown in cross-section, with the two branches of the obturator nerve sandwiched between the layers

5 The *sartorius* is a long strap-like muscle arising from the anterior superior spine of the ilium and traversing the front of the thigh superficially downward and inward to its insertion in the upper tibia. It helps flex both hip and knee and to sit with crossed legs, the 'tailor's muscle'.

6 The *hamstrings* are the bulk of the back of the thigh, a powerful group arising from the ischial tuberosity which flex the knee by inserting into the tibia and fibula. They are the *semimembranosus*, *semitendinosus* and *biceps femoris*, which diverge as they pass down the thigh. The first two pass inwards to the back of the upper tibia, while the biceps femoris, which acquires an additional head from the back of the femur, is inserted on the outer side of the knee into the head of the fibula. This Λ-shaped arrangement of the tendons produces the upper boundary of the ◇-shaped *popliteal fossa* at the back of the knee, the lower sides of the diamond being the two heads of the gastrocnemius muscle. The *sciatic nerve*, which lies in the thigh under cover of the hamstrings, emerges between the upper fork to lie comparatively superficially at the apex of the popliteal region. The space is roofed by deep fascia and has the popliteal surface of the femur as its floor. The disposition of these muscle groups is shown in cross-section (Figures 7.24, 7.25). The intermuscular septa of deep fascia demarcate three compartments, each containing a muscle group with its supplying nerve, which also innervates the overlying skin.

▶ The *anterior* or extensor compartment contains the quadriceps, supplied by the femoral nerve.

▶ The *medial* contains the adductors, supplied by the obturator nerve.

▶ The *posterior* contains the hamstrings, supplied by the sciatic.

The sartorius in its diagonal course is applied to the side of the vastus medialis – to produce a tunnel-like space called the *adductor canal*, through which pass the femoral vessels in their journey round the inner side of the femur to emerge in the popliteal fossa as the popliteal artery and vein.

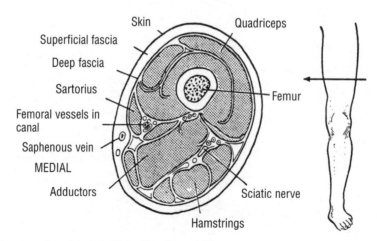

Figure 7.24 Cross-section of the left thigh, viewed from below, to show the muscle masses

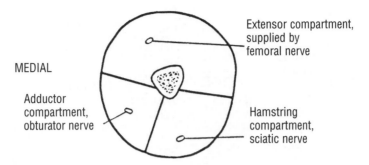

Figure 7.25 Key to Figure 7.24: cross-section, viewed from above, of the three compartments of the thigh, with their nerves

Muscles of the leg

Cross-section of the leg (Figures 7.26, 7.27) reveals the main compartments and muscle groups as follows. The interosseous membrane between tibia and fibula separates an *anterior compartment* containing the muscles that extend the ankle and toes from the *posterior compartment* for the calf muscles. There is a small separate *lateral compartment* on the outer side of the fibula for the peroneal muscles which evert the foot. The main posterior compartment is subdivided ito a *superficial* portion for the gastrocnemius and soleus, and a *deep* portion for the long flexors of the toes and the tibialis posterior which inverts the foot.

ANTERIOR GROUP

This group consists of the tibialis anterior, extensor hallucis longus and extensor digitorum longus. They are extensor muscles, supplied by the anterior tibial division of the peroneal nerve. Their tendons run in front of the ankle, where they are bound down by retinacular bands of deep fascia. The *tibialis anterior*, on the medial side, is inserted into the first metatarsocuneiform junction, where it helps maintain the inner long arch and dorsiflexes the foot. The extensors of the toes and the separate tendon for the great toe (hallux) traverse the dorsum of the foot to be inserted like the corresponding tendons in the hand, except that in the foot there is a duplicate set of extensor tendons arising from a short extensor muscle (*extensor digitorum brevis*) situated on the dorsum of the foot itself.

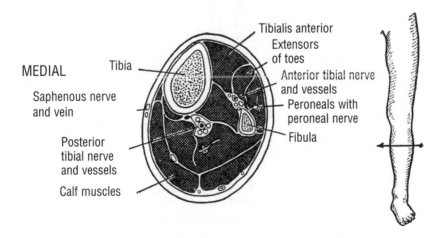

Figure 7.26 Cross-section of left lower leg, seen from below, showing the main muscle masses

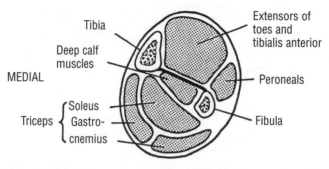

Figure 7.27 Key to Figure 7.26: cross-section, viewed from above, of compartments of the lower leg

The *peroneal* muscles, *longus* and *brevis*, evert the foot and dorsiflex the ankle; they are supplied by the peroneal nerve. The brevis has a short course to the base of the fifth metatarsal, whereas the longus dives under the sole and runs across to reach the base of the first metatarsal, so helping maintain the transverse arch. Both tendons pass behind the lateral malleolus to reach the foot.

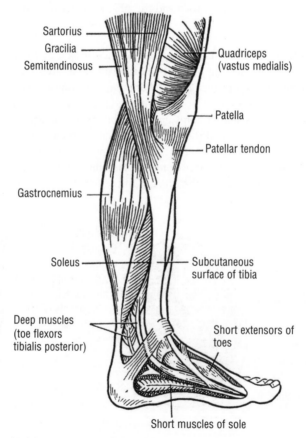

Figure 7.28 Muscles of left lower leg and foot, inner aspect, medial view

POSTERIOR GROUP

1 *Superficial layer*. This comprises the superficial *gastrocnemius*, with medial and lateral heads arising from the back of each femoral condyle, and the underlying *soleus* arising from the back of the upper tibia and fibula. These join to form the tendon halfway down the calf. These muscles plantarflex the foot and ankle through its insertion via the Achilles tendon into the tuberosity of the calcaneus.

2 *Deep layer: tibialis posterior, long flexors of toes*. These arise from the back of both bones and interosseous membrane. Their tendons enter the sole by passing behind the ankle and medial malleolus. The long flexors are attached to the toes (with a separate muscle for the great toe). The tendons of the calf correspond only to the profundus, and the duplicate tendons are supplied by a short muscle in the sole, the *flexor digitorum brevis*. The *tibialis posterior* is inserted into the navicular, sustains the inner long arch by taking strain off the spring ligament by supporting the head of the talus, and is a powerful invertor of the foot as well as a plantarflexor of the ankle.

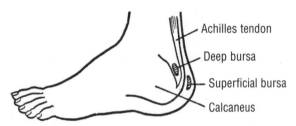

Figure 7.29 Achilles tendon and related bursae

FOOT

The general arrangement of the foot and toes, flexor and extensor tendons, their synovial sheaths and so on, resembles that of the hand, with local differences due to the duplication of the long flexor and extensor tendons by short *intrinsic* extensors and flexors situated entirely in the foot itself, arising from the upper and under surfaces of the calcaneus.

The deep plantar fascia of the sole is very dense and covers a complex layered pattern of small plantar muscles.

Blood vessels of the lower limb

The *femoral artery* and *vein* are the distal continuations of the *external iliac* vessels from the pelvis. They pass under the inguinal ligament, carrying a sheath prolonged from the deep fascia of the abdomen, and lie in the groin with the vein medial to the artery. Their course in the thigh lies in three parts: in the upper third, in the inguinal region; in the middle third, in the adductor canal, under the sartorius muscle; in the lower third they pass round to the back of the femur to enter the popliteal fossa as the *popliteal* vessels. In the thigh the artery supplies the muscle masses, and it divides in the popliteal fossa into an *anterior* and *posterior tibial* branch. The posterior tibial continues the line of the vessel into the calf, where it lies close to the interosseous membrane, enters the sole behind the medial malleolus with the flexor tendons and supplies *digital arteries* to the toes. The anterior tibial passes forward from the popliteal fossa through the interosseous membrane into the anterior compartment of the leg and runs down the front of the membrane to emerge on the dorsum of the foot and give branches to the toes. Because the femoral artery is so superficial at the groin, it is useful for insertion of a catheter for injection of radio-opaque agents to outline the circulatory system under X-rays, especially the coronary arteries of the heart.

The main venous channels are similar.

Nerves of the lower limb

The *femoral* and *sciatic* nerves are the great anterior and posterior trunks and, with the *obturator* and certain other twigs, are derived from the *lumbosacral plexus* of spinal nerve roots situated in the abdomen and pelvis.

The *femoral nerve* enters the thigh beneath the inguinal ligament, lateral to the femoral vessels, and breaks up into the following branches:

1 the lateral, intermediate and medial cutaneous nerves of the thigh

2 muscular branches to the quadriceps femoris

3 densory twigs to the hip and knee joints.

The *obturator nerve* enters the adductor comparment of the thigh from the pelvis by piercing the obturator membrane; it supplies the adductors, the overlying skin and the knee joint.

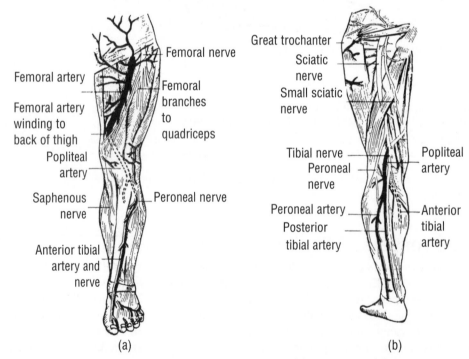

Figure 7.30 Plan of the main nerves and vessels of the lower limb: (a) anterior view; (b) posterior view

The *sciatic nerve* is formed in the pelvis, on the inner aspect of the sacrum, emerges into the buttock, where it lies deeply under the gluteal muscles, and runs vertically down to the politeal fossa, where it divides into medial (*tibial)* and lateral (*peroneal*) branches. In the upper part of the thigh it lies between ischial tuberosity and great trochanter; in the mid-thigh it is covered by the hamstrings and becomes more superficial when the hamstring tendons diverge to their insertions.

The *tibial nerve* continues the course of the sciatic from the upper to the lower angle of the popliteal fossa, lying superficially under the deep fascia and crossing over the popliteal vessels. It supplies the calf muscles, forming a common *neurovascular bundle* with the posterior tibial artery, with which it enters the foot as the *medial* and *lateral plantar* nerves to supply the intrinsic muscles and sensory branches to the toes.

The *peroneal nerve* follows the biceps femoris tendon to the head of the fibula, winds superficially round the neck of that bone, gives a superficial branch to the peroneal muscles and the skin of the outer side of the calf, and continues as the *anterior tibial nerve* with the anterior tibial artery in the extensor compartment. Here it supplies the extensor muscles, enters the dorsum of the foot and gives digital branches to the toes.

Note that here, as elsewhere in the body, main axial vessels – arteries and their corresponding veins and lymphatics – form common neurovascular bundles with the nerves, with a surrounding sheath. The larger arteries also often have a pair or plexus of their own small satellite veins, the *venae comitantes*, closely hugging their walls.

Test yourself

1 Which area of the inominate bone has the socket for the femoral head?
  a   Ishium
  b   Ilium
  c   Pubis
  d   Acetabulum

2 Which area on the femur is the lowest attachment for the adductor muscles?
  a   Adductor tubercle
  b   Poplitela fossa
  c   Interconylar notch
  d   Linea aspera

3 Which area of the tibia overhangs the inner side of the ankle.
  a   Medial malleolus
  b   Tibial tuberosity
  c   Lateral malleolus
  d   Medial condyle

4 Identify the bone that is the heel bone.
  a   Tarsals
  b   Metatarsals
  c   Phalanges
  d   Calcaneus

5 Which ligaments link the femoral condyles to their opposite tibial condyles heel bone?
  a   Collateral
  b   Spring
  c   Cruciate
  d   Inguinal

6 When the foot is inverted, what other type of movement accompanies this?
  a   Adduction
  b   Abduction
  c   Rotation
  d   Flexion

7 Identify the location of the bursae in the lower limb that may be inflamed by excessive cycling.
  a   Between the skin and ischial tuberosity
  b   Over the outer aspect of the great trochanter
  c   In front of the patellar tendon
  d   Between the skin and Achilles tendon

8 Which of the following muscles is part of the gluteal muscle group?
  a   Longus
  b   Medius
  c   Brevis
  d   Magnus

9 Identify the muscle that helps a person to sit with their legs crossed.
   a Gracillia
   b Psoas
   c Semimembranosus
   d Sartorius

10 Which muscle contracts to evert the foot?
   a Peroneal brevis
   b Tibialis anterior
   c Extensor digitorum brevis
   d Gastrocnemius

8

Regional anatomy: the abdomen

In this chapter you will learn about:

▶ *the abdominal cavity and its boundaries, surface anatomy and muscles*

▶ *the peritoneum, mesenteries and omenta*

▶ *the general disposition of the abdominal organs*

▶ *the liver*

▶ *the spleen*

▶ *the posterior abdominal wall and related structures*

▶ *the pelvis*

▶ *the bladder*

▶ *the rectum and anus*

▶ *other pelvic structures, including vessels and nerves.*

Abdominal cavity and boundaries

The abdominal cavity is the largest body space; its roof, the diaphragm, reaches high into the chest under the lower ribs, and its lowest part, the pelvic cavity, is partly enclosed within the pelvic bones. The right and left domes of the diaphragm separate the right and left lungs above from the corresponding lobes of the liver below, with the heart and pericardium sitting on the flat middle area of the partition. In full expiration the right diaphragm is at the level of the fifth rib, and the left is 2.5 cm (1 inch) lower.

The *abdomen proper* contains the main internal organs – bowel, liver, pancreas, spleen, kidneys, adrenals and great vessels – and the *pelvic cavity* below, enclosed by the innominate bones, contains the terminal bowel, bladder and genital organs. The two cavities are continuous at the *pelvic inlet*. A longitudinal section through the trunk shows that the main axes of the two cavities are at an angle, with backward inclination of the pelvis, whose organs are relatively separate from those in the abdomen proper.

The abdomen is relatively unprotected by bony framework in front and at the sides, though this is compensated to some extent by a strong layered muscle wall.

The boundaries of the abdomen are:

1 *behind*, the lumbar vertebrae of the spinal column, clothed by the psoas and quadratus lumborum muscles

2 *in front and at the sides*, the muscles of the flank and anterior abdominal wall

3 *above*, the diaphragm

4 *below*, on each side, the iliac fossae of the innominate bones, clothed with the iliacus muscles, which support part of the abdominal contents.

The cavity is lined with a serous membrane, the *peritoneum* (see below), which also clothes most of the contained organs, or *viscera*, some solid, some hollow. All these peritoneal surfaces are normally in contact and the cavity is only potential unless air is admitted by operation or injury to create a *pneumoperitoneum*. A cross-section shows how the vertebral column encroaches forward so that it can be felt through the anterior abdominal wall, leaving a bay on either side in which lie the kidneys (see Figure 8.2).

There is considerable respiratory excursion in the abdomen. On inspiration, the diaphragm is depressed as the lungs expand and depresses the abdominal organs, the liver descending 5–8 cm (2–3 inches) with a deep breath in. The cavity also varies in size with contraction and relaxation of its muscular walls and the degree of distension of the hollow viscera.

Surface anatomy

ANTERIOR ABDOMINAL WALL

Above, on each side, the *costal margins*, the lower borders of the ribs, form a Λ-shaped angle. At its apex is the tip of the *xiphoid process* of the sternum. The *umbilicus* or navel is in the midline, on a level with the fourth lumbar vertebra. In the lowest part of the midline anteriorly is the *pubic symphysis*, the hard fibrous junction between the two halves of the pelvis in front. Xiphisternum, umbilicus and symphysis are connected by a tough ribbon of deep fascia, the *linea alba* or white line, a firm central attachment for the muscles, whose aponeuroses intersect here. When they contract the linea alba is seen as a central depression between the *rectus abdominis* muscle on either side; the belly of the latter is crossed by two or three intersections. All this is well seen in body-builders.

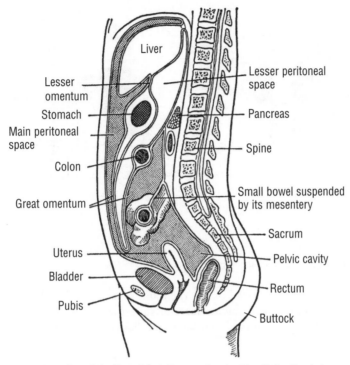

Figure 8.1 Longitudinal section of the (female) abdomen, showing the distinction between pelvic cavity and abdomen proper

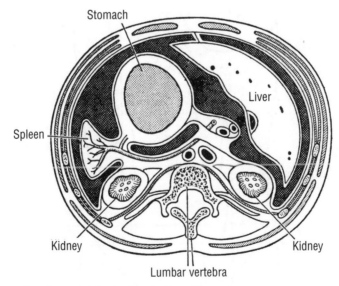

Figure 8.2 Cross-section of upper abdominal cavity, viewed from above

Note the recess on either side of the vertebral column in which each kidney lies

The large black space is the greater sac of the peritoneal cavity; the lesser sac is the small space immediately behind the stomach

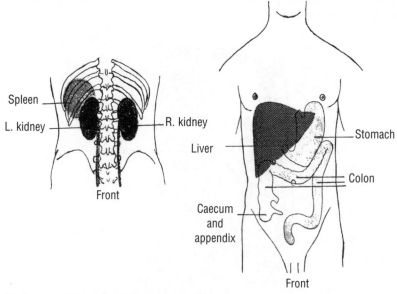

Figure 8.3 Surface markings of certain abdominal organs

Inferiorly the symphysis can be traced out to the pubic crest on each side; the *inguinal ligament*, spanning from pubic tubercle to anterior superior iliac spine, marks the junction between abdomen and thigh. From the iliac spine the *iliac crest* can be followed round to the back.

The surface of the anterior abdominal wall is divided into zones for reference (Figure 8.4). In the upper zone is the *epigastric* region centrally, with the *hypochondriac* region on either side.

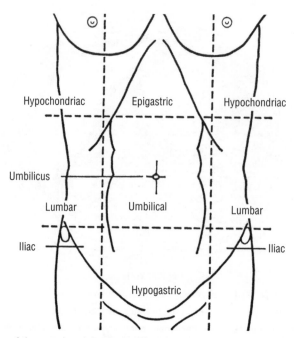

Figure 8.4 The zones of the anterior abdominal wall

In the middle zone the umbilical region is central, flanked by the *lumbar* regions. Below, the *hypogastric* region is central and the *iliac* regions to right and left.

POSTERIOR ABDOMINAL WALL

In the midline of the back the *spinous processes* of the lumbar vertebrae are felt between the thorax and the sacrum; and on each side of the spine is the belly of the great *erector spinae* muscle. The short *twelfth rib* on each side marks the lowest limit of the thoracic cage, while below the iliac crests can be traced back to the sacrum, ending in the *posterior superior iliac spine*, marked by a dimple in the overlying skin. On each side of the spinal column, beyond the lateral border of the *erector spinae* muscle, is an uprotected portion of the abdominal wall, the loin or *lumbar* region, between twelfth rib and iliac crest, overlying the kidneys. However, the flank musculature is extremely strong.

Abdominal muscles

These are more easily drawn than described and fall into the following groups:

1   the muscles of the posterior abdominal wall – *psoas, quadratus lumborum*

2   the muscles of the flanks – *internal* and *external obliques, transversus abdominis*

3   the muscles of the anterior abdominal wall – *rectus abdominis* and the small triangular *pyramidalis* below.

The cross-section in Figure 8.5 gives a general idea of the relations of these groups.

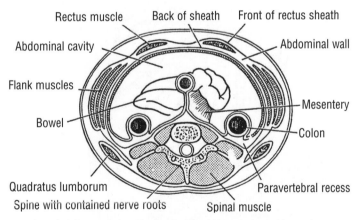

Figure 8.5 Cross-section of abdomen below the level of the kidneys, viewed from above

Note the suspension of the small bowel by its mesentery from the posterior abdominal wall; also the relations of the muscle groups

MUSCLES OF POSTERIOR ABDOMINAL WALL

The *psoas* has already been encountered in connection with the hip joint (Chapter 7); a long belly beside the lumbar vertebrae, it skirts the pelvic brim to enter the thigh with a contribution from the iliacus. The *quadratus lumborum* lies immediately lateral to it, a quadrilateral sheet between twelfth rib and iliac crest.

FLANK MUSCLES

These are arranged in layers, with the *internal oblique* sandwiched between the *external oblique* superficially and the *transversus abdominis* deeply. Their fibres cross in different directions, adding to their protective power. They arise from the lower ribs above, the iliac crest below and, via a dense sheet of lumbar fascia, from the tips of the lumbar transverse processes behind. The external oblique fibres run down and in, like a person putting their hand in their pocket. The internal oblique fibres are at right angles to this, and those of the transversus run straight round.

Near their origins above, below and behind, these muscles are very fleshy, but as they curve round the flank to the anterior abdominal wall they merge into broad *aponeurotic sheets* extending between costal margin and iliac crests. These aponeuroses are intimately related to each other and to the rectus abdominis muscle, as indicated below, they intersect at the *linea alba* at the midline a fibrous band extending from the xiphoid to the pubis.

THE APONEUROSES, RECTUS SHEATH, NERVES AND VESSELS OF THE ANTERIOR ABDOMINAL WALL

About halfway between the flank and the linea alba, the flank muscles continue forwards as aponeurotic sheets. As they approach the lateral border of the rectus muscle the external oblique aponeurosis passes in front of that muscle and the transversus aponeurosis behind, while that of the internal oblique splits, one layer joining each of the others (Figure 8.6). This provides a continuous *rectus sheath*, whose anterior and posterior walls rejoin on the inner side of the rectus to form the linea alba.

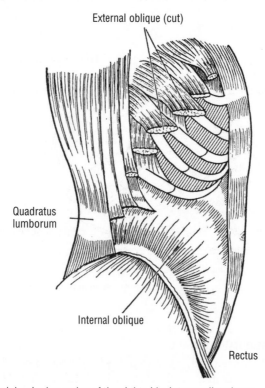

External oblique (cut)

Quadratus
lumborum

Internal oblique

Rectus

Figure 8.6 The flank abdominal muscles of the right side, intermediate layer

The external oblique has been removed, and the internal oblique is seen extending between the quadratus lumborum behind and the rectus abdominis in front

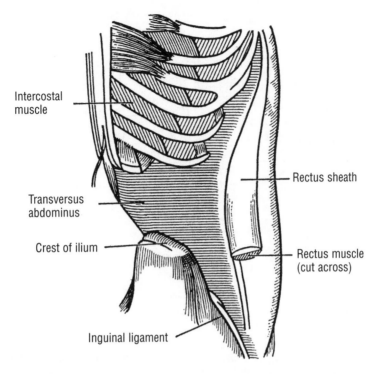

Figure 8.7 The deepest layer of the flank muscles, transversus abdominis, lateral view (after *Gray*)

The *nerves and vessels* of the anterior abdominal wall are the lower six pairs of thoracic nerves and vessels. As the lower ribs overhang the abdominal cavity and are incomplete anteriorly (because they do not reach the sternum), the neurovascular bundles in the rib spaces emerge betwen the flank muscles. They run between the transversus and internal oblique to the lateral border of the rectus, enter its sheath behind the muscle, and finally turn forward at its inner border, piercing the linea alba to reach the skin. In their circular course, branches are given off to the muscle layers.

Finally, within the rectus sheath a vertical arrangement of arteries and veins is formed behind the rectus muscle by anastomosis between the *superior epigastric* vessels descending from the thorax and the *inferior epigastrics* running up from the external iliac trunks below (Figure 8.8).

Peritoneum, mesenteries and omenta

The abdominal cavity is a closed sac, lined with the serous glistening peritoneal membrane. The membrane clothing the deep aspect of the abdominal wall is the *parietal* peritoneum; that reflected over the contained viscera is the *visceral* peritoneum. The parietal layer is attached to the deep surface of the anterior and posterior abdominal walls, the underside of the diaphragm and the upper surface of the pelvic floor. Its smooth surface allows the structures to glide freely, and a loose layer of connective tissue intervenes between the parietal peritoneum and the abdominal wall.

The main peritoneal cavity of abdomen and pelvis is known as the *greater sac*, contrasted with a smaller recess, the *lesser sac*, which lies behind the stomach. These spaces, and the relations

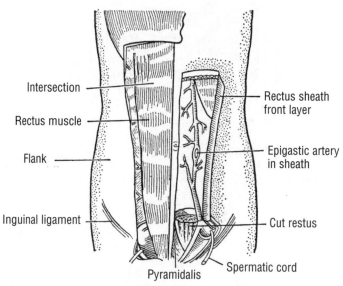

Intersection

Rectus muscle

Flank

Inguinal ligament

Rectus sheath
front layer

Epigastic artery
in sheath

Cut restus

Pyramidalis

Spermatic cord

Figure 8.8 The left rectus sheath opened and the muscle removed to show the epigastric vessels, anterior view

of the organs to the peritoneum, are complex and best understood by studying Figures 8.1 and 8.2. Most of the organs lying in the abdominal cavity have developed in the embryo from the posterior wall, to which they still retain an attachment. These are the main points (Figure 8.9).

1 Certain organs are entirely retroperitoneal and therefore fixed, e.g. the kidneys and pancreas; they lie on the posterior abdominal wall with a peritoneal covering, if any, on their anterior surfaces only.

2 Other organs, the ascending and descending parts of the colon and the rectum – parts of the large bowel – have a peritoneal covering in front and at the sides; they are still retroperitoneal, but with more mobility.

3 The small intestine and the transverse and pelvic portions of the colon hang freely; to do so, they have pulled out a double-layered sheet of peritoneum from the posterior abdominal wall, which is known as a *mesentery*. This is laden with fat, lymph nodes and vessels, and the blood vessels reach the bowel by running between its layers from the great vessels on the posterior abdominal wall. Small bowel, transverse and sigmoid colon each have their own mesentery; but the main one is the great sheet supporting the small bowel, known simply as *the* mesentery, whose line of attachment crosses the lumbar spine obliquely from left to right and above downwards (Figure 8.5).

4 The stomach has two special mesenteries known as *omenta*. The *great omentum* hangs down from its lower border as an apron-like fold in front of the small bowel and then turns up to embrace the transverse colon before its final reflection onto the posterior abdominal wall, i.e. the mesentery of the transverse colon is really continuous with the great omentum of the stomach (Figures 5.8 and 8.1). The *lesser omentum* connects the upper border of the stomach to the liver and forms the anterior boundary of the lesser sac (Figure 8.1).

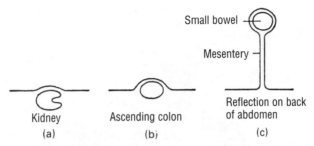

Figure 8.9 Different degrees of peritonization of abdominal organs: (a) a retroperitoneal organ; (b) organ with peritoneum on three sides; (c) freely suspended organ with mesentery

Nugget

The mnemonic SAD PUCKOR can be used to help to remember which organs are retroperitoneal in the body:

S suprarenal (adrenal) gland

A aorta

D duodenum (second and third part)

P pancreas (except tail)

U ureters

C colon (ascending and descending)

K kidneys

O oesophagus

R rectum

The general disposition of the abdominal organs

Figure 5.8 shows the structures revealed when the anterior abdominal wall is removed. The *liver* occupies the upper right portion of the cavity, emerging a little below the costal margin, and a little on the left side, with its intermediate portion lying in the epigastric region. Beneath its lower border, on the right side, projects the *gall-bladder*. The *stomach* is largely under cover of the left ribs, but part of its anterior surface is seen in the angle between the lower edge of the liver and the left costal margin. The *great omentum* descends from the lower border of the stomach, covering the *transverse colon*, which lies immediately below, and the coils of the *small intestine*. These last are the main contents of the cavity, insinuating themselves into its recesses and descending into the pelvis. The *spleen* is barely seen unless enlarged by disease, as it is tucked under the left costal margin. The more superficial viscera are now to be considered.

STOMACH

The stomach is the widest part of the digestive tract, connecting the *oesophagus* (gullet) with the *duodenum*, the beginning of the small intestine. The oeophagus joins the stomach shortly after passing through the diaphragm into the abdomen, and the duodenum leaves the organ at the *pyloric orifice*. The stomach lies in the epigastric and left hypochondriac regions, but there is considerable variation with posture, digestive state and emotion. It is J-shaped, with

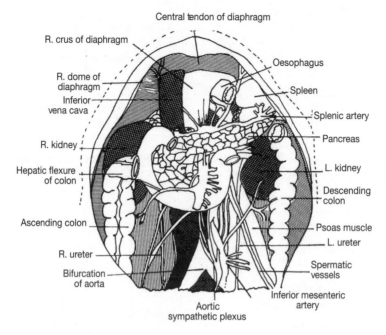

Figure 8.10 Structures of the posterior abdominal wall after removal of the liver, stomach, small intestine and transverse colon, viewed from above

The duodenum, which is unlabelled, is the C-shaped loop of bowel in the centre

The transverse colon normally arches across it to connect the ascending and descending limbs of the large bowel, here shown cut across

anterior and posterior surfaces, and upper and lower borders known as the *lesser* and *greater curvatures* respectively. The lesser curve sweeps from the oesophageal entry at the *cardia* on the left extremity to the departure of the duodenum at the *pylorus* at the right-hand end.

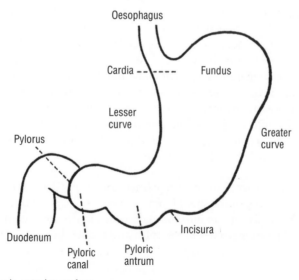

Figure 8.11 The stomach, anterior surface

The main subdivisions are:

1 the dome-shaped *fundus*, the receptive portion, often distended with air

2 the main *body*, concerned with digestion, marked off by a definite notch or *incisura*

3 the *pyloric* or expulsive portion, narrowing as the *pyloric antrum* to the *pyloric canal*.

Both cardiac and pyloric orifices are surrounded by *sphincters*, muscular rings that are normally contracted to keep the openings closed until relaxation is required for entry or exit of food.

The *stomach* is made up of several layers (Figure 8.12):

▶ the smooth outer *serous* or peritoneal coat

▶ the intermediate *muscular* wall, which contains circular, longitudinal and oblique fibres

▶ the internal lining *mucous membrane*, a velvety folded layer whose glands secrete the gastric juice.

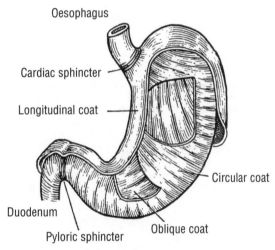

Figure 8.12 The muscle coats and sphincters of the stomach, lateral view

The *anterior surface* of the stomach is under cover of the left lobe of the liver and left costal margin, with a small intermediate portion just behind the anterior abdominal wall. The *posterior surface* lies on a 'stomach bed' (Figure 8.13), formed by the organs of the posterior abdominal wall – pancreas, left kidney and suprararenal, and spleen. Between the posterior surface and its bed is the *lesser sac*. The *fundus* lies in contact with the undersurface of the left dome of the diaphragm.

SMALL INTESTINE
This is that part of the digestive tract between the pyloric orifice and the large bowel. Some 6 m (20 feet) long, it is divided into:

1 the *duodenum*, a short loop of 30 cm (12 inches) immediately continuous with the stomach and bound down tightly to the posterior abdominal wall

2 the *small intestine proper*, suspended in coils from the posterior wall by its mesentery.

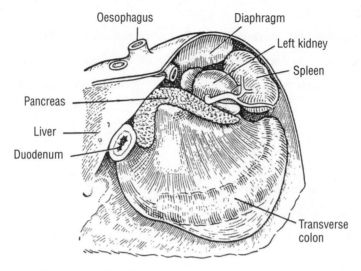

Figure 8.13 The stomach bed, viewed from below

The left lobe of the liver has been cut away, and the stomach itself removed, to show the organs on which it lies (after *Gray*)

The *duodenum* is C-shaped, embracing the head of the pancreas.

▶ A short *first part* runs horizontally from the pylorus under cover of the liver and gall-bladder.

▶ The vertical *second part* descends in front of the right kidney.

▶ The *third part* crosses to the left in front of the lumbar vertebrae, from which it is separated by the great vessels – aorta and inferior vena cava. It joins the small intestine at

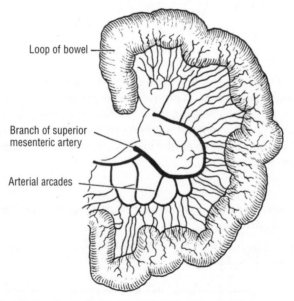

Figure 8.14 A loop of small bowel, with its blood vessels, viewed from above

the *duodenojejunal flexure*. The pancreatic and bile ducts from the liver have a common opening into its second part.

The *small intestine* runs from the duodenojejunal flexure to the *ileocolic valve*, which marks its junction with the *caecum*, or first part of the large bowel. Its coils are completely peritonized, except for a narrow strip, the *mesenteric border*, where the two layers of the mesentery diverge to enclose them. The small bowel resembles the other parts of the intestine in possessing an outer *serous* coat, a *muscular wall* and a lining *mucosa*. But characteristic are the *circular folds* projecting into its lumen. The enormous number of minute fringes or *villi* of the mucosa give it a velvety appearance like the pile of a carpet. These act as a device for increasing the absorptive area and the scattered patches of *lymphoid tissue*. Networks of blood and lymph vessels form plexuses between the layers, all forming larger trunks which run into the mesentery. The blood vessels of the bowel enter the root of the mesentery as the *superior mesenteric* branches of the aorta and inferior vena cava, and form a series of arcades between its layers, from which the ultimate twigs are given off to the bowel.

The *jejunum* is the upper two-fifths of the small bowel, the *ileum* the lower three-fifths. The ileum is thinner, narrower, less vascular and contains more lymphoid tissue, as digestion and secretion preponderate at the upper end of the bowel and absorption at its lower. The small gut as a whole is enclosed by the framework of the limbs of the colon. The jejunum is above and to the left, the ileum central and below, some loops spilling into the pelvis before rising again as the terminal ileum to join the large bowel.

THE MESENTERY

The leaves of the mesentery enclose:

1 fat and connective tissue

2 lymph nodes receiving the bowel lymphatics or *lacteal* vessels

3 the mesenteric vessels

4 nerves supplying the intestine.

The posterior attachment or 'root' of the mesentery is an oblique line of some 15 cm (6 inches) running down the posterior abdominal wall from the left of the second lumbar vertebra to the right sacroiliac joint below, crossing the duodenum and great vessels. The great disparity between this short origin and its extensive attachment to the bowel causes the membrane to be thrown into fan-shaped folds.

LARGE BOWEL

The large intestine (Figure 8.15) runs from the end of the ileum to the external orifice of the anus. Only some 1.5–1.8 m (5–6 feet) long, it includes:

1 the caecum, with its vermiform appendix

2 the ascending, transverse and descending limbs of the colon

3 the pelvic colon

4 the rectum and anal canal.

Only the transverse and pelvic colon have mesenteries and are freely mobile; the rest is retroperitoneal. The lower rectum and anus are entirely below the level of the peritoneum, in the depths of the pelvis.

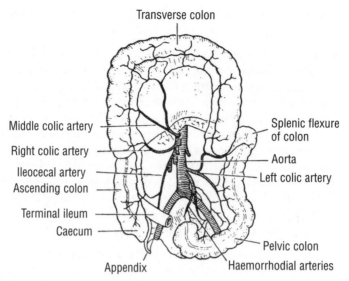

Figure 8.15 The large bowel and its blood vessels

The transverse loop of colon, which is normally dependent, has been turned up for clarity

The *caecum* lies in the right iliac fossa, the hollow basin formed by the iliac portion of the right innominate bone. It is the blind sacculated commencement of the large bowel, the terminal ileum opening into it by an aperture guarded by the *ileocaecal valve*. The appendix is attached a little lower in the angle between. The caecum has no mesentery but is completely invested by peritoneum, and has the mobility of a flapping balloon.

The *appendix* is a worm-like tube, blind at its free end, and with a tiny mesentery of its own. Its position is far from constant; it may be behind the caecum, hang over the pelvic brim, or turn up behind the terminal ileum. Probably a vestigial organ, its main importance is as a cause of surgical emergency.

COLON

The *ascending*, *transverse* and *descending colon* are arranged round the abdominal cavity like three sides of a square – Π – but the two vertical limbs lie in the paravertebral gutter on each side and therefore behind the plane of the transverse limb, which sags down towards the pelvis. The junctions of transverse colon with the vertical limbs at either side are known as flexures, the *hepatic flexure* on the right under cover of the liver, and the *splenic flexure* on the left in relation to the spleen. Because of the greater bulk of the liver on the right side, the hepatic flexure is a few centimetres lower than the splenic. The colon generally is distinguished by:

1 three superficial bands of longitudinal muscle standing out under the serous coat and traversing the bowel from end to end – the *taenia coli*, which are spaced equally round its circumference

2 the scattered fatty tags or polyps which project as the stalked *epiploic appendages* on the surface

3 segmentation into coarse sacculations or *haustrations*, vaguely resembling those of an earthworm and marked internally by *semilunar folds*.

The internal structure is like that of the bowel in general, but the serous coat is incomplete where the colon is only partly peritonized and the mucosa is smooth and pale without villi.

The *ascending colon* lies on the posterior abdominal wall, mainly on the quadratus lumborum muscle and is relatively immobile.

The *hepatic flexure* lies in front of the lower pole of the right kidney, overhung by the right lobe of the liver.

The *transverse colon* arches across the abdomen, lying below the stomach and obscured by the great omentum. It is very mobile as it has its own mesentery, the *transverse mesocolon*, derived from the doubling back of the layers of the great omentum, which diverge to enclose the bowel and rejoin to form its mesentery as they pass to the posterior abdominal wall (Figure 8.1).

The transverse colon lies in front of most of the structures of the posterior abdominal wall at this level – duodenum, pancreas, great vessels and portions of the kidneys on either side – and the prominence of the lumbar spine behind arches the colon forwards.

The *splenic flexure* lies in front of the left kidney, immediately below the spleen, with the diaphragm behind.

The *descending colon*, like the ascending, has no mesentery and is plastered onto the posterior abdominal wall behind the peritoneum. It runs down the left side from the splenic flexure as far as the pelvic brim, where the shallow left iliac fossa gives way to the true deep pelvis, crossing in front of the left psoas muscle and left common iliac artery and vein.

The *pelvic colon*, the continuation of the descending colon, extends from the pelvic brim to the beginning of the rectum, opposite the middle of the sacrum in the depths of the true pelvis. It has a long mesentery or *pelvic mesocolon* and is loose and mobile.

The *rectum* and *anus* will be described in connection with the pelvis (see below).

BLOOD VESSELS OF THE LARGE BOWEL (FIGURE 8.15)

Each colonic limb has its main artery: the ascending and transverse colon have the *right* and *middle colic* vessels – branches of the *superior mesenteric* artery, which has already supplied the small bowel – while the descending limb receives the *left colic* division of another branch of the aorta, the *inferior mesenteric* artery, which passes down to the rectum as the *superior rectal* artery after giving twigs to the pelvic colon. The right, middle and left colic arteries approach the bowel, fork into two main branches which run parallel with it to anastomose (interconnect) with adjacent vessels. They thus form a continuous arterial channel whose branches form a network whose ultimate twigs reach the intestine. The pattern of the veins is similar; but, whereas the arteries have sprung from the aorta, the veins do not return to the companion inferior vena cava but enter an entirely separate system behind the pancreas, the great *portal venous trunk* which gathers up the splenic vein to run in the lesser omentum and enters the liver. This so-called portal circulation ensures that the venous blood from the bowel, containing the products of digestion, passes through the liver for processing before entering the general circulation and returning to the heart. This is why all medication by mouth is less effective than administration by injection under the skin or into a limb vein.

LIVER

This is a bulky solid organ – the largest gland in the body – in the upper right part of the abdominal cavity. It has a brownish friable substance, with a smooth peritoneal coat, and is slung under the diaphragm by *suspensory ligaments* which are the peritoneal reflections.

The larger *right lobe* and the smaller *left lobe* lie beneath the right and left domes of the diaphragm, and most of the organ is under cover of the ribs. It reaches almost to nipple level on each side and projects slightly beyond the inferior costal margin in the epigastric and right hypochondriac regions, where it touches the back of the anterior abdominal wall.

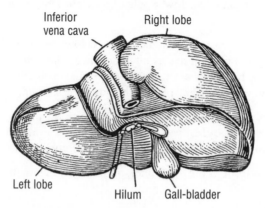

Figure 8.16 The liver, seen from behind (after *Gray*)

Because of the upward bulge it produces in the floor of the thoracic cavity, the lower borders of the lungs and pleural cavity actually surround the upper part of the liver. The organ is ⊽-shaped in sagittal section, the sloping limb facing backwards, with superior, anterior and posterior surfaces, and bluntly rounded borders of which the inferior is the sharpest.

The *anterior surface* is behind the lower ribs and anterior abdominal wall in the epigastric region. Demarcation of right and left lobes is by the *falcifom ligament,* which attaches the liver to the back of the anterior abdominal wall in the midline between xiphisternum and umbilicus. The *superior surface* rests against the diaphragm. The *posterior surface* is more complex. On the *right* it overlies the right kidney and hepatic flexure of the colon; on the *left* it lies on the stomach, upper pole of the left kidney and suprarenal gland, and the spleen.

The *gall-bladder* is attached to the back of the right lobe and can be seen from in front projecting below its lower border. The *inferior vena cava* is embedded in the back of this lobe on its way to pierce the diaphragm and enter the thorax.

At the centre of the back of the liver, between the lobes, is the root or *hilum,* where the main vessels and ducts enter and leave the organ. These are:

1 the *hepatic artery* from the aorta, dividing into right and left branches to the two lobes

2 the *portal vein,* carrying digested nutriment from the bowel, also dividing into two branches

3 the right and left *hepatic ducts,* conveying bile from each lobe and joining to form the *common hepatic duct.* The *cystic duct* from the gall-bladder joins the common duct a little below its formation and the main channel so formed is the *common bile duct.*

These structures – bile duct, hepatic artery and portal vein – run between the layers of the lesser omentum, which is attached to the hilum of the liver at one end and the lesser curve of the stomach at the other.

The functions of the liver include processing of carbohydrate (see Chapter 15) and proteins for utilization by the body after absorption from the bowel, the storage of carbohydrate as glycogen, and secretion of bile. The latter is stored in the gall-bladder, a reservoir without secretory function of its own. It has a muscular coat which contracts in response to the entry of food into the duodenum. The bile empties into the duodenum through the common duct, which joins with the duct from the pancreas in a joint *ampulla* just before entering the bowel.

Within the bowel the bile emulsifies the fat of food for easier digestion.

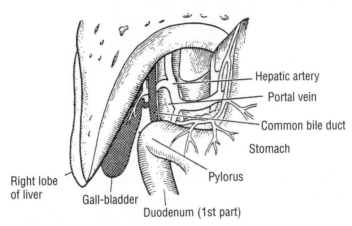

Figure 8.17 The liver, seen from in front, with part of the right lobe cut away to show the structures entering and leaving the hilum (after *Gray*)

SPLEEN

This lies just below the left dome of the diaphragm, corresponding to the right lobe of the liver on the opposite side. It is a much smaller organ, entirely under cover of the ribs, soft and pulpy, with a fibrous capsule. It is pyramidal or tetrahedral, with a large smooth convex surface under the diaphragm and three smaller surfaces facing inwards and converging on the *hilum*, where the splenic artery and veins are attached. One surface forms part of the stomach bed (Figure 8.13); the others are in contact with the left kidney and splenic flexure of the colon. The tail of the pancreas reaches to the hilum.

The spleen is a reservoir of red blood corpuscles. There are muscle fibres in its capsule and it can squeeze more cells into the circulation if effort or lack of oxygen require. It is also a site for the formation of the lymphocytes of the blood and immune bodies.

It is vulnerable to rupture from violent compression, as in traffic accidents, causing severe haemorrhage which requires emergency removal of the organ to save life. It may also be removed for lymphomatous disorders such as Hodgkin's disease. After such a splenectomy, the body's defences against infection are reduced.

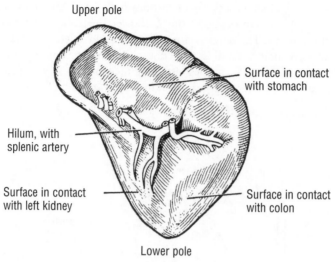

Upper pole

Surface in contact
with stomach

Hilum, with
splenic artery

Surface in contact
with left kidney

Surface in contact
with colon

Lower pole

Figure 8.18 The spleen, anterior view

Posterior abdominal wall and related structures

Looking into the abdominal cavity from the front after removal of the small bowel, we see the *posterior abdominal wall*, a bony/muscular boundary partly obscured by certain structures. These are the *pancreas* and *duodenum*, stretching from side to side; the *kidneys* and their *ureters*, one in each paravertebral recess; and the *great vessels* running downward in the midline.

The *posterior wall* is composed in the midline of the bodies of the lumbar vertebrae, which project forward leaving a *paravertebral recesses* on either side in which lie the kidneys and the ascending or descending colon (Figure 8.2). On each side of the bodies is the *psoas* muscle, and, more laterally, the flat *quadratus lumborum* muscle, between twelfth rib above and iliac crest below. The plexus of lumbar nerve roots which emerge between the vertebrae lies embedded in the psoas and its branches supply the lower limb. The upper limit of the posterior wall is the lowest attachment of the diaphragm behind; its lower limit is the pelvic brim; laterally, it is continuous with the flank muscles.

GREAT VESSELS: AORTA AND INFERIOR VENA CAVA

The *abdominal aorta* is the continuation through the *aortic aperture* of the diaphragm, at the level of twelfth thoracic vertebra, of the thoracic portion of this great artery, which has arisen from the left ventricle of the heart. It runs down in the midline in front of the lumbar vertebral bodies as far as the fourth lumbar vertebra, where it divides into the *right* and *left common iliac arteries* at a level just below and to the left of the umbilicus. It therefore lies very deeply (though its pulsation can be felt from in front in a thin person) and is crossed by the pancreas and the third part of the duodenum. There is a mass of sympathetic nerve tissue on each side of its commencement, the *coeliac ganglia* or *solar plexus*, branches of which form an *aortic plexus* around the vessel. The main *sympathetic nerve chain* lies on the vertebral bodies on either side.

The *inferior vena cava* is a great vein lying immediately to the right of the aorta, draining blood from the lower abdomen and lower limbs. It is formed by the union of the two *common iliac veins* in front of the fifth lumbar vertebra, the junction lying behind the right common iliac artery. It runs up the posterior abdominal wall to pierce the diaphragm and enter the right atrium of the heart.

The branches of the aorta include:

1 single unpaired branches in the midline – the *coeliac axis* to liver, spleen and stomach, the *superior mesenteric* to small bowel, ascending and transverse colon, and the *inferior mesenteric artery* to the remainder of the large bowel

2 symmetric paired branches on each side to the diaphragm, suprarenals, kidneys, testicles or ovaries. The largest are the *renal arteries*; the right renal has to pass behind the vena cava to reach its kidney, and since the ovaries lie in the pelvis, and the testicles even lower, their vessels have to run down on the posterior abdominal wall in company with the ureters.

The branches forming the *inferior vena cava* correspond to those of the aorta, with adjustment for the difference in position, e.g. the left renal vein is much longer than the right and has to cross in front of the aorta. But the venous blood from the bowel has to pass through the liver before entering the general circulation and this is achieved by a secondary *portal circulation*. The *superior* and *inferior mesenteric veins* join behind the pancreas to form the main *portal vein*, which also drains the stomach and spleen. This trunk ascends in the lesser omentum to the liver, receiving a branch from either lobe. In the omentum it forms a common bundle wth the bile duct and hepatic artery. After the portal blood has passed through the liver capillaries, it re-enters the general venous stream by a number of small *hepatic veins* opening directly into the vena cava at the back of the liver.

The main *lymphatic* drainage from the abdomen, including the *lacteal vessels* carrying digested fat from the bowel, is into a small receptacle, the *cisterna chyli*, lying between the upper parts of aorta and vena cava. It enters the chest with the aorta to become the *thoracic duct*, of matchstick size, which travels up to the root of the neck on the left side to empty into the great veins.

PANCREAS

This is a soft, solid, lobulated organ of ⊂-shape embraced in the concavity of the duodenum and stretching across the posterior abdominal wall and great vessels. It is composed of a *head*, the rounded right extremity, grasped in the duodenal curve; a *neck*; the *body*, which stretches across to the left side in front of the great vessels and the upper part of the left kidney; and a *tail*, turning up to the hilum of the spleen.

The pancreas has important external and internal secretions. The *external secretion*, the *pancreatic juice*, aids digestion and is discharged through the pancreatic duct into the second part of the duodenum. At its termination this duct forms a common channel with the common bile duct so that bile and pancreatic juice are discharged simultaneously as required; the opening is closed between meals by a sphincter.

The *internal secretion*, from island groups of cells, is *insulin*, which passes into the bloodstream and is essential for tissue utilization of sugar. Deficiency of this secretion is the cause of diabetes (see also Chapter 15).

The superior mesenteric vessels emerge from behind the body of the pancreas and then pass in front of its head to enter the mesentery of the small bowel. The splenic artery runs along its upper border from the coeliac branch of the aorta to the spleen. The lower parts of portal vein, hepatic artery and common bile duct are situated in or behind the head and neck.

KIDNEYS AND URETERS

The kidneys remove waste products and excess water from the blood brought to them by the renal arteries, and the urine they excrete is passed into ducts, the *ureters*, which carry it to the bladder (Figure 5.11). Each kidney lies in a paravertebral recess. They are bean-shaped organs, convex at their outer borders and concave towards the midline, with a body and upper and lower poles. Perched on the upper pole is the *suprarenal gland*, an organ of internal secretion.

They lie obliquely, the upper pole nearer the midline. The right kidney is 2.5 cm (1 inch) or more lower than the left, owing to the bulk of the right lobe of the liver. The left kidney rests on the eleventh and twelfth ribs, but the right only on the twelfth. At the middle of the concave medial border is a depression, the root or *hilum*, where the artery, vein and ureter are attached.

The kidney has a glistening *true capsule* and lies embedded in a voluminous *false capsule* of perirenal fat in which it glides up and down with respiration. Since the kidneys lie on the lower ribs, the lowest part of the pleural cavities of the thorax overhang the back of the upper pole, separated by the lowest fibres of the diaphragm, but most of the organ lies on the psoas and quadratus muscles. The other relations differ on the two sides.

▶ *Right*: the right lobe of the liver is in front of most of the anterior surface; the second part of the duodenum lies before the hilum; the hepatic flexure of the colon is in front of the lower pole.

▶ *Left*: the body of the pancreas crosses the middle of the anterior surface, which also forms part of the stomach bed. The spleen is applied to the convex outer border, and the splenic flexure of the colon is in contact with the lower pole.

Longitudinal section (Figure 8.19) shows an outer solid substance enclosing an inner cavity, the *renal pelvis*, which collects the urine. The renal substance has an outer rind or *cortex* and a deeper *medulla*, and the latter is arranged in pyramidal masses whose apices project into little bays of the pelvis as the *calyces*. The microscopic renal tubules in which the urine is formed discharge at the tips of the pyramids into the pelvis. The latter is partly enclosed within the kidney but protrudes at the hilum to become continuous with the ureter.

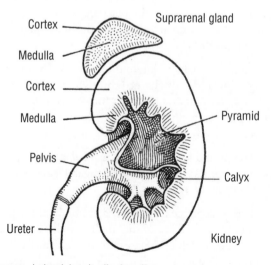

Figure 8.19 Kidney and suprarenal gland, longitudinal section

The *ureters* begin at the *pelviureteric junction* in the hilum, emerge from behind the renal vessels, and run down and in on the psoas muscle beind the peritoneum. At the pelvic brim they cross in front of the common iliac vessels, run down the side-wall of the true pelvis and enter the bladder. They are hollow muscular tubes, some 25 cm (10 inches) long and 4 mm (1/5 inch) in diameter, down which urine is propelled in spurts by waves of contraction.

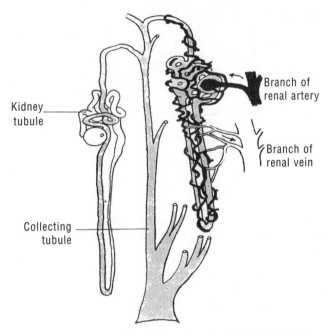

Figure 8.20 Two kidney nephrons – the microscopic coiled tubes in which the urine is formed – draining into a common collecting tubule

The blood vessels which carry blood to and from each tubule for purification are shown on the right

THE DIAPHRAGM

This is the chief muscle of respiration and forms a partition, part muscular, part tendinous, between thoracic and abdominal cavities. It has a right and left muscular *dome* rising high into the thorax and separating the lungs from the abdominal viscera, and an intermediate flat *central tendon* on which the heart rests.

The muscular portion has a number of bony origins. In *front*, it arises from the back of the xiphistenum, at the *sides* from the lower ribs; and in the midline *posteriorly* it helps form the upper part of the posterior abdominal wall by arising as two pillars or *crura* from the sides of the upper three lumbar vertebrae (Figure 8.21).

There are several apertures for structures passing between the thorax and abdomen. The *aortic aperture* is embraced by the two crura as they cross in front of the twelfth thoracic vertebra. The *oesophageal aperture* is in the left dome at the level of the tenth thoracic body, and the opening for the *inferior vena cava* lies in the central tendon to the right of the midline at the level of the ninth thoracic vertebral body.

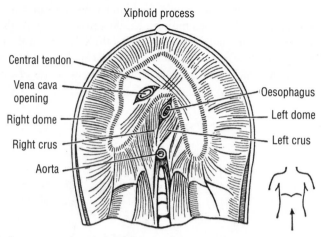

Central tendon

Vena cava opening

Right dome

Right crus

Aorta

Xiphoid process

Oesophagus

Left dome

Left crus

Figure 8.21 The diaphragm, seen from below (after *Gray*)

Pelvis

The lower portion of the abdominal cavity is known as the pelvic cavity and is inclined backward to form rather a separate compartment, although the two are continuous at the pelvic inlet (Figure 8.1).

The cavity is formed by the basin-shaped arrangement of the bony pelvis: the two innominate bones at the sides and in front and the sacrum and coccyx behind (Figures 3.13 and 3.14). The bony pelvis falls into two parts (Figures 8.23 and 8.26):

1 the *false pelvis* or *pelvis major*, the shallow surfaces of the basin above the pelvic brim, i.e. the *right* and *left iliac fossae*, clothed by the iliacus muscles and supporting the caecum and pelvic colon respectively

2 the *true pelvis* or *pelvis minor*, the deeply enclosed portion below the pelvic brim containing the essential pelvic organs.

The true pelvis is what is usually meant by the word, and the following remarks apply to it alone. It has an *inlet*, an *outlet*, a *cavity* and a *floor*. Its boundaries are only partly bony and are completed in front by soft tissues, and the floor is entirely made of soft tissues: the *levator ani* muscle and fascia.

The *inlet* faces forward and slightly upward, due to the backward inclination of the sacrum. Its boundary is the *pelvic brim*, seen from above as heart-shaped, with the projecting upper part or *promontory* of the sacrum encroaching behind and the angle of the pubic arch meeting at the symphysis in front (Figures 3.13 and 3.14).

The *cavity* is a conical canal. Its walls are:

▶ *anteriorly*, the back of the pubic symphysis and rami

▶ *laterally*, the innominate bones, with the obturator foramen bridged by the obturator membrane

▶ *posteriorly*, the anterior surface of sacrum and coccyx.

These bony walls are incomplete and are lined by muscles, pelvic fascia, and by peritoneum in its upper part. The chief contents are the pelvic (sigmoid) colon and rectum on the posterior wall, in the hollow of the sacrum, and the bladder in front, behind the symphysis. In women the uterus and vagina are interposed between rectum and bladder; in men the bladder rests on the prostate gland and seminal vesicles.

The *outlet* is a diamond-shaped space seen from below. The four bony angles of the ◊ are: *anteriorly*, the symphysis; at each *side* the ischial tuberosity; and *posteriorly* the tip of the cocccyx. The two *anterior limbs* of the ◊ are bony, the pubic arch formed by the inferior pubic rami running up to the symphysis; the two *posterior limbs* are the *sacrotuberous ligaments* spanning from sacrum and coccyx to the ischial tuberosities.

SEX DIFFERENCES IN THE BONY PELVIS

As the *female pelvis* has to allow the passage of the baby's head, it is roomier than the male and shallower. The female pelvis has its side-walls more vertical; the iliac fossae are shallower; the inlet is large and nearly circular; the sacrum is short and wide and only projects slightly into the cavity; the outlet is wide, the pubic arch making an obtuse angle, while the coccyx is very mobile.

The *male pelvis* is a narrow cavity with sloping walls, a small heart-shaped inlet with marked sacral encroachment, a tighter outlet with the pubic arch forming a right angle, and a more rigid coccyx.

SURFACE ANATOMY

The only region where the pelvis comes near the surface is at the *perineum*, the space between the thighs which contains the orifices of the genital, urinary and digestive tracts. Figure 8.22 shows the female perineal structures. There is a diamond-shaped area of peroneal skin divided into two triangles by a line joining the ischial tuberosities. The posterior *anal triangle* contains the opening of the *anus*, continuous with the rectum above. The anterior *urogenital triangle* contains the opening of the *vagina*, the lower end of the genital tract, with the *urethra* just

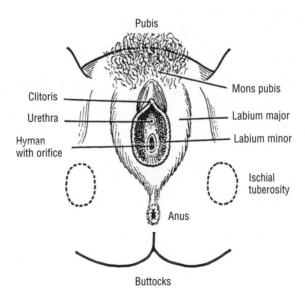

Figure 8.22 The female perineum

in front, the urinary channel which runs from the bladder through the pelvic floor; further forward is the *clitoris*.

In men the anal triangle is similar, but the urogenital region is different. The penis has at its base a central *bulb*, attached to the midpoint of the perineum, into which the urethra passes from the bladder through the prostate and pelvic floor (Figure 8.24); and it has two supporting *crura*, attached to the pubic arch on either side and converging on the bulb to form the *shaft* of the organ, which is traversed by the *urethral canal*, opening at the external urinary meatus at its tip.

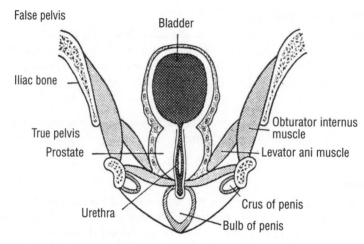

Figure 8.23 Coronal section of male pelvis

Note the distinction between true and false pelvis; also the muscles of the pelvic floor (after *Gray*)

THE PELVIC FLOOR (LEVATOR ANI MUSCLE)

The pelvis is like a funnel with a very wide stem; the false pelvis forms the sloping sides and the true pelvis the vertical portion. The pelvic organs are contained in the vertical portion and rest on its *floor*, a musculofascial partition slung from the side-walls of the cavity. This embraces the various canals – anal, vaginal and urethral – which pierce it to reach their perineal orifices, surrounding each with a sphincteric grip. The upper aspect of the pelvic floor is lined by the parietal layer of peritoneum, which is reflected over the contained organs. The general arrangement is shown in longitudinal section in Figures 8.24 and 8.25, and is considerably simpler in the male.

In the *male* the peritoneum is reflected from the back of the abdominal wall over the upper surface of the bladder, dips down between bladder and rectum (as the *rectovesical pouch*) and turns up again over the front of the rectum. Most of the bladder is below peritoneal level, the prostate entirely so.

In the *female* the peritoneum, after covering the bladder, is thrust upward by the projection of the uterus; it covers the front and back of that organ, passing off the latter to form the *rectouterine pouch* – the most dependent part of the abdominal cavity – before ascending again on the rectum.

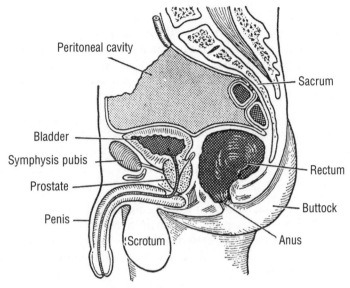

Figure 8.24 Midline sagittal section of the male pelvis (after *Gray*)

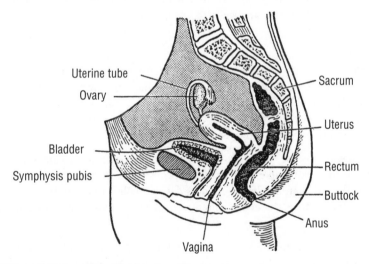

Figure 8.25 Midline sagittal section of the female pelvis

In both sexes the lower limit of peritoneum is the third piece of the sacrum. The upper part of the rectum has a peritoneal covering in front but no mesentery; the lower part is entirely below peritoneal level.

PELVIC VISCERA

Bladder and rectum are the same in both sexes; the genital organs – uterus, uterine tubes, ovaries, vagina, prostate, seminal vesicles, testicles and their ducts – are described in Chapter 22.

▶ Bladder

This is a hollow muscular organ in the front of the pelvic cavity. It receives urine via the ureters and expels it into the urethra by micturition. Its shape varies with distension, but is roughly that of an inverted pyramid (Figure 22.3). When empty, it lies entirely within the pelvis behind the pubis. As it distends, its domed upper part or *fundus* ascends into the abdomen in contact with the back of the anterior abdominal wall.

Its upper surface is covered by peritoneum and the ureters open into its upper lateral angles by oblique passages through the muscular wall, at the *uretic orifices*. The tapering, dependent bladder neck is continuous with the urethra at the *internal urinary meatus*. The two ureteric orifices and the meatus form the three points of the *trigone*, a region of the bladder base very sensitive to irritation.

In *women* the organ is overhung from behind by the uterus, and the urethra is a canal of only 4 cm (1 1/2 inches) which pierces the pelvic floor to open immediately in front of the vagina. In *men* the bladder neck rests on the *prostate gland*, into which the male urethra passes; the *seminal vesicles* for storage of sperm and the ducts conveying sperm from the testicles are close to the lower part of the bladder.

The bladder wall has a thick muscle coat and a mucosal inner lining, which is wrinkled when the organ is contracted. The involuntary nervous system adjusts muscle tone to fluid content, so that it relaxes as it fills, keeping the tension constant until a threshold is reached when the tension is felt as an urge to micturate (see also Chapter 19).

▶ Rectum and anus

The rectum is the lowest part of the large bowel; faeces enter from the pelvic colon and remain until discharged by defaecation. It is some 13 cm (5–6 inches) long, begins as the continuation of the pelvic colon at the middle of the sacrum, follows the hollow of the sacral curve, and turns forward at the coccyx to join the anal canal – a short wide passage 4 cm (1 1/2 inches) long that bends sharply backward to open at the *anal orifice*.

The rectum has a wavy course, and its lower portion or *ampulla* is capable of considerable distension. Two or three valve-like mucosal folds project into its lumen. Only the upper portion has a peritoneal coat, on the front and sides. Its wall is longitudinal and circular muscle, and the lowest part is thickened as a powerful ring gripping the anorectal junction, the *internal sphincter*.

The anal canal is distensible for the passage of faeces; its upper part is lined with mucosa, but the lower part by inturned skin continuous with that of the perineum. Immediately under the perianal skin encircling the orifice is another circular muscle, the *external sphincter*. Both sphincters are normally closed. In the act of defaecation they relax, the abdominal wall tightens and raises abdominal pressure, and the rectum is lifted up over the contained faeces by the muscular pelvic floor.

OTHER PELVIC STRUCTURES

The *ureters* cross the pelvic brim, passing over the common iliac vessels, travel down the side-walls of the pelvis under the peritoneum 5 cm (2 inches) from the rectum, and sweep inward and forward to reach the bladder.

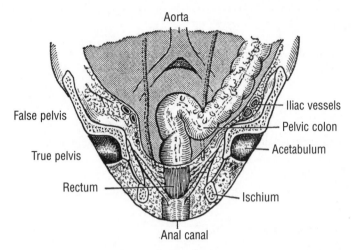

Aorta

Iliac vessels

Pelvic colon

Acetabulum

False pelvis

True pelvis

Rectum

Ischium

Anal canal

Figure 8.26 Coronal section of pelvis to show the rectum and anus; note the distinction between true and false pelvis; also the lower limit of the peritoneal cul-de-sac (after *Gray*)

▶ **Vessels**

The *common iliac* vessels pass downward and out from their origin at the bifurcations of the aorta and inferior vena cava and divide at the level of the lumbosacral joint into *external* and *internal iliac* branches. The external iliacs continue along the pelvic brim together with the psoas tendon and femoral nerve, and pass beneath the inguinal ligament to emerge as the femoral vessels in the groin. The internal iliacs descend the side-walls of the true pelvis and branch to supply the bladder, rectum and genital organs.

▶ **Nerves**

The *lumbar plexus* of nerves on the posterior abdominal wall is succeeded in the pelvis by the *sacral plexus* of nerve roots, emerging from the foramina of the sacrum and lying on the back of the cavity. The main branch of the lumbar plexus is the *femoral nerve*, which accompanies the psoas tendon and external iliac vessels to the groin to supply the muscles and skin of the front of the thigh; the main branch of the sacral plexus is the *sciatic nerve*, which leaves the back of the pelvis to emerge in the buttock.

The *sympathetic nerve chain* continues down from the abdomen on the front of the sacrum and coccyx, ending as a single fused ganglion on the latter.

Test yourself

1 Which zone sits at the lowest point in the abdominal wall?
   a Hypochondriac
   b Epigastric
   c Lumbar
   d Hypogastric

2 Which abdominal muscles are located at the posterior of the abdominal wall?
   a Internal obliques
   b Quadratus lumborum
   c Rectus abdominis
   d Pyramidalis

3 Identify the lowest part of the stomach.
   a Fundus
   b Incisura
   c Pyloric antrum
   d Pyloric canal

4 Which organs lie on the posterior abdominal wall with a peritoneal covering?
   a Small intestine
   b Descending colon
   c Kidneys
   d Rectum

5 Which organ is located closest to the liver?
   a Pancreas
   b Spleen
   c Kidney
   d Gall-bladder

6 Where in the stomach is the fundus located?
   a Undersurface of the diaphragm
   b Posterior surface of the liver
   c Anterior surface of the pancreas
   d Undersurface of the duodenum

7 How long is the short loop of the duodenum?
   a 10 cm
   b 20 cm
   c 30 cm
   d 40 cm

8 Identify the shape of the duodenum.
   a C-shaped
   b S-shaped
   c J-shaped
   d L-shaped

9 What structures increase the surface area of the small intestine?
   a Villi
   b Ileocolic valve
   c Mesenteric border
   d Serous coat

10 What is the main branch of the lumbar plexus?
   a Sympathetic nerve chain
   b Sciatic nerve
   c Femoral nerve
   d Sacral plexus

9

Regional anatomy: the thorax

In this chapter you will learn about:

▶ *the bony wall and ribs*

▶ *respiration*

▶ *the muscles of the chest wall*

▶ *the thoracic cavity and its contents – trachea and bronchi, lungs, and heart*

▶ *the great vessels of the thorax*

▶ *the aorta*

▶ *other structures of the posterior thoracic wall*

▶ *the nerves within the thorax.*

The thorax is a bony cage for the heart and lungs, It appears to be cylindrical, but without the shoulders (Figure 9.1) it is seen to be collical or barrel-shaped, with a narrow apex.

Bony wall

In the midline behind are the twelve *thoracic vertebrae* and in the midline anteriorly the breastbone or *sternum*. Between them are the encircling *ribs*. Note in cross-section (Figure 9.4) the *paravertebral recess* on each side of the spine due to the projection of the vertebral bodies and the backward curve of the ribs before they turn forward.

RIBS

There are twelve pairs of ribs, separated by *intercostal spaces* which contain the intercostal muscles, nerves and vessels.

▶ The first seven are *true ribs*, complete from spine to sternum.

▶ The eighth, ninth and tenth are *false ribs*, turning up to join the ribs above.

▶ The eleventh and twelfth are short *floating ribs*, embedded at their tips in the flank muscles of the abdomen.

The first and twelfth ribs are very short, the longest the seventh and eighth, so that the barrel contour of the chest slopes in again above and below. Each rib also slopes somewhat downwards.

Each rib has:

▶ a *head*, articulating with the side of a vertebral body

▶ a short *neck*, lying on the transverse process of the vertebra (Figure 3.15(b))

▶ a *body* which curves back a little and then sweeps sharply forward at the *angle*

▶ the *costal cartilage*, a gristly rod of 2.5–5 cm (1–2 inches) connecting rib to sternum. The framework is deficient below in front, where the *costal arch* is formed by the right and left *costal margins*.

The *intercostal spaces* are filled by layers of muscles, between which the *intercostal vessels* and *nerves* encircle the chest wall. One nerve, artery and vein occupy each space, a good example of the persistence of primitive segmentation.

The *sternum* or breastbone is dagger-shaped, with three main components. Uppermost is the broad, flat *manubrium*, to which are attached the inner ends of the clavicles and first ribs. Below is the main *body*, a flat bone of two compact layers sandwiching spongy bone with red marrow, with the costal cartilages attached on either side. Lowest is the small pointed *xiphisternum*, in the upper abdominal wall.

The thorax has an *inlet* above, through which the great vessels and nerves pass to and from the neck, or over the first rib into the axilla and arm. The inlet is narrow and closely packed, only 6.5 cm (21/2 inches) separating the manubrium from the spine, with the first ribs on either side. The most important structures traversing the inlet are the *oesophagus*, *trachea*, certain *nerves* and the *great arteries* and *veins*. The thorax is sealed off below by the diaphragm.

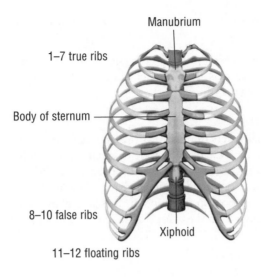

Figure 9.1 The bony thoracic cage, seen from in front

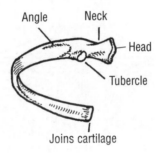

Figure 9.2 A typical rib, seen from behind

Respiration

Respiration draws air into and pushes it out of the lungs. In *inspiration*, the chest cavity is enlarged and air enters; in *expiration* the reverse occurs. Respiratory movements are both thoracic and abdominal. In *thoracic* inspiration the sternum is lifted by elevation of the ribs, which come to lie more horizontally. Both the transverse and anteroposterior diameters of the chest are increased. In *abdominal* inspiration, the diaphragm contracts and flattens, pressing the abdominal organs down and bulging the abdominal wall; the vertical height of the thorax is increased. Inspiration is an active process, due to muscular exertion. Expiration is passive: the chest wall subsides, the abdominal wall recoils, the diaphragm relaxes and air is driven out of the lungs.

Muscles of the chest wall

ANTERIOR

In front, the chest wall is covered by the great *pectoralis major*, mentioned in connection with the arm (Figure 6.10(a)). Arising from sternum, clavicle and ribs, its fibres converge to be inserted into the upper humerus. Its function is adduction of the arm to the side. A smaller *pectoralis minor* lies beneath it, arising from a few ribs and inserted into the coracoid process of the scapula.

Both pectorals can act as *accessory muscles of respiration* if the arm is fixed. In the lower chest wall some of the abdominal muscles are attached to the ribs: the *rectus* near the midline and the *oblique* muscles more laterally.

POSTERIOR (FIGURE 9.3)

The two great superficial sheets of muscle at the back are the trapezius above and the latissimus dorsi lower down. Each is a muscle originating from the spine and controlling the shoulder girdle. The *trapezius* is triangular, with its base attached to the spinous processes of the thoracic vertebrae and extending up the back of the neck as far as the occiput. It narrows rapidly to its insertion on the spine of the scapula and back of the clavicle. It rotates the scapula clockwise on the chest wall (Figure 6.10) and thus aids the deltoid in abducting the shoulder.

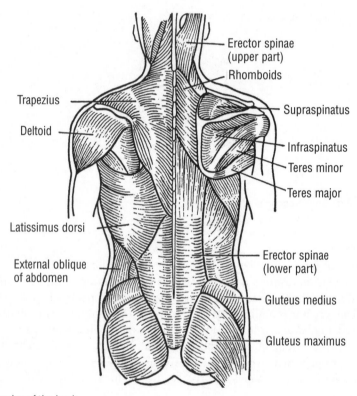

Figure 9.3 Muscles of the back

On the right side the superficial muscle layer has been removed to expose the deeper structures

The *latissimus dorsi* is overlapped by the trapezius above. It originates from the lower thoracic vertebrae and lumbar spine, via a dense sheet of fascia. Its fibres pass away and up to insert into the humerus at the same level as the pectoralis major. It adducts the arm. Placing one's fingers in the armpit, a thick muscle boundary is felt in front and behind: the pectoralis and latissimus respectively.

Thoracic cavity

The thoracic cavity is divided into right and left halves by a massive partition, the *mediastinum*, which lies in the sagittal midline, stretching from the back of the sternum to the vertebral column. Vertically, the cavity extends from thoracic inlet to diaphragm. The two halves are separate and contain the right and left lungs. Each lung is attached to the mediastinum at its root or *hilum* by its bronchus (airway) and blood vessels.

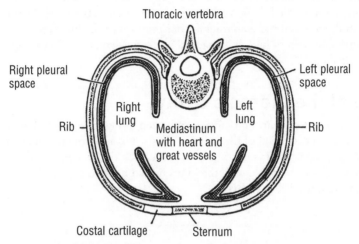

Figure 9.4 Cross-section of the chest to show its division into right and left halves by the mediastinal partition, viewed from above

The cavity is lined by a smooth serous membrane, the *pleura*, which facilitates gliding of the lung on the chest wall. Each pleural sac is a closed sac, for the pleura, like the peritoneum, is divided into parietal and visceral layers which are everywhere continuous. The *parietal pleura* clothes the deep aspect of the ribs, the upper surface of the diaphragm and the sides of the mediastinum. The *visceral layer* ensheaths the lung, the two layers being continuous at its root.

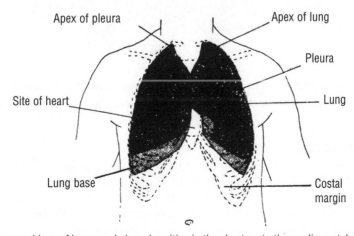

Figure 9.5 Surface markings of lungs and pleural cavities in the chest: note the cardiac notch of the left lung

The apex of the pleural sac rises into the root of the neck, a little above the clavicle, and its base overhangs the liver in front and the kidneys behind. Although the lung closely fills the pleura, it does not quite reach its upper and lower limits. Normally, there is no actual pleural cavity, for the layers are in contact and the lung fills its side of the thorax. The lung is elastic and tends to shrink and expel its contained air. This cannot occur normally as a negative pressure exists in the potential pleural space. If air is admitted, the lung immediately shrinks into a small solid mass, collapsed against the mediastinum.

The *heart* is embedded in the mediastinum, occupying a central position between the lungs, though extending more to the left. It is enclosed in a fibrous sac, the *pericardium*, which rests on the central tendon of the diaphragm. After removal of the lungs, the intervening mediastinal partition is seen to be solidly packed with vital structures – heart, great vessels, trachea and oesophagus.

Contents of the thoracic cavity

The general arrangement of heart and lungs is indicated in Figure 5.8, and Figure 9.6 shows the thoracic viscera after removal of the anterior chest wall, and how the lungs so overlap the heart that only a small part of it is in contact with the back of the sternum. Figure 9.6 shows the organs removed from the chest, with the lungs turned back to expose the great vessels and heart; the pericardium has been removed.

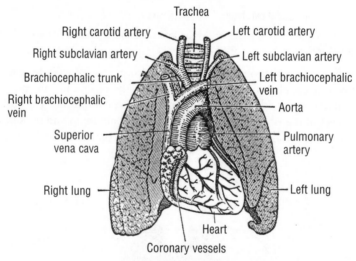

Figure 9.6 Lungs, heart and great vessels, seen from in front after removal from the chest
The lungs normally overlap the greater part of the heart and have been retracted to expose that organ (after *Gray*)

TRACHEA AND BRONCHI

The *trachea* enters the chest from the neck at the thoracic inlet and runs down to the sternal angle, where it divides into a *right* and *left main bronchus* for each lung (Figure 5.10). In its short course it has the oesophagus behind, separating it from the spine, and the great vessels – aorta and superior vena cava – in front, separating it from the back of the sternum. The bronchi enter the lung hila, together with the pulmonary vessels. All these structures emerge from the mediastinum to reach the lungs. Trachea and bronchi have numerous cartilaginous rings in their walls to keep them from collapsing and maintain a free airway. In the trachea these rings are deficient behind, so that the windpipe is semicircular in section (Figure 10.8).

LUNGS

The lungs are light spongy organs designed to provide the maximum surface area for the interchange of oxygen and carbon dioxide between air and blood. Each lung is divided into main *upper* and *lower lobes* by an oblique fissure running down and forward; but the right lung has a third or *intermediate lobe* marked off from the upper lobe by a short transverse fissure. The lungs are permeated by elastic connective tissue and, when removed from the chest, collapse to a quarter of their expanded size. The organ, in whole or part, floats in water.

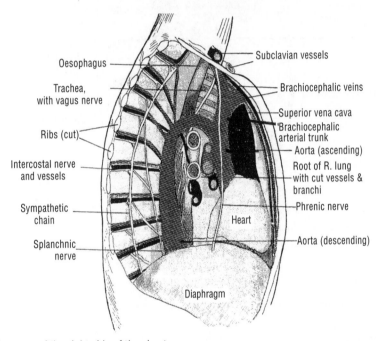

Figure 9.7 Contents of the right side of the chest
The lung and pleura have been removed to exhibit the mediastinal structures

The lungs lie free in the chest enclosed in the visceral pleura, attached only by their hila to the mediastinum. Each *apex* rises into the base of the neck some 1–3 cm (1/3–1 inch) above the middle of the clavicle. The *base* rests on the upper surface of the diaphragm, which separates it on the right from the right lobe of the liver and on the left from the left lobe, stomach and spleen. The *outer surface* is in contact with the inner aspect of the chest wall and is indented by the ribs. The rounded *posterior border* fits into the recess at the side of the thoracic vertebrae and the *anterior border* overlaps the heart and pericardium. Because the bulk of the heart is on the left side there is a well-marked *cardiac notch* in the anterior border of the left lung.

The *medial surface* faces inwards towards the mediastinum and bears the hilum. The relations of the medial surfaces differ on the two sides, since the mediastinal structures are not symmetric. The main differences are due to:

1 the curvature of the arch of the aorta over the left bronchus to the left side and the prominence of the descending aorta on the left side

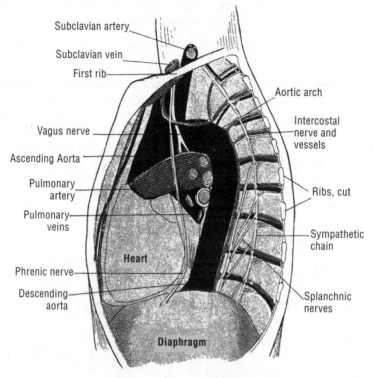

Figure 9.8 Contents of the left side of the chest

The lung and pleura have been removed to exhibit the mediastinal structures

2 the presence of the superior vena cava on the right

3 the right-sided position of the oesophagus below.

▶ Internal structure of the lungs

Each main bronchus divides into a branch for each lobe and these subdivide into a ramifying tree of fine *bronchioles*, accompanied by branches of the pulmonary artery and vein. Each bronchiole ends in a cluster of *alveoli*, where interchange occurs between the contained air and gases dissolved in the blood of the perialveolar capillaries. Each lung *lobule* consists of a terminal bronchiole and its alveoli, and the lung has millions of lobules bound together by elastic connective tissue and providing an interchange area the size of a tennis court.

HEART

The heart is a hollow muscular organ lying between the lungs in the mediastinum. It is enclosed in the *pericardium*. This membrane has a tough outer *fibrous layer*, firmly blended below with the central tendon of the diaphragm, and a delicate inner *serous layer*. The serous pericardium lines the fibrous capsule and is reflected at the roots of the great vessels springing from the heart to cover the organ. The pericardial sac is unyielding, and if haemorrhage occurs within it due to a stab wound of the heart, the pressure rises so greatly as to fatally compromise cardiac function unless early surgery is available.

The heart has four hollow chambers: the two *atria* above, which receive blood from the great veins, and the two *ventricles* below, which expel it into the great arteries. Although the atrium and ventricle of the same side can communicate freely, but in a strictly one-way direction from atrium to ventricle, they are shut off from the opposite pair by a muscular *septum*, which is much thicker between the ventricles than the atria. The latter are thin-walled compared wth the fleshy ventricles, and the wall of the left ventricle is thicker than the right. On the surface, grooves indicate the demarcation between the chambers, and in these grooves run the important *coronary vessels* which nourish the heart muscle (Figure 9.9).

The shape of the heart is roughly conical, with the apex directed to the left and slightly downwards. The anterior surface – behind the sternum and costal cartilages – is fomed mainly by the right ventricle with a little of the right atrium and left ventricle on either side. The right border is entirely right atrium; the left border is left ventricle and the apex is at its tip. The left atrium lies entirely on the posterior surface, facing the vertebral column, oesophagus and descending aorta. The uppermost tips of the atria are little ear-like extensions, or *auricles*.

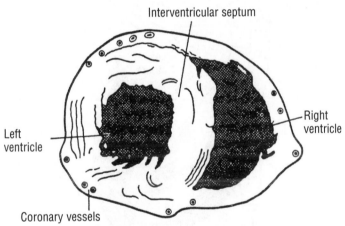

Figure 9.9 Cross-section of heart in ventricular region, viewed from above; note the greater thickness of the left ventricle, the massive septum and the coronary vessels cut across

ATTACHMENTS OF THE GREAT VESSELS

Two great veins enter the right atrium: the *superior vena cava* above and the *inferior vena cava* below. The venous blood flows through the chamber to the right ventricle and is expelled into the pulmonary artery arising from its upper border, going to the lungs in the two branches of this vessel. At the back of the heart four *pulmonary veins* return oxygenated blood from the lungs to the left atrium, which passed it to the left ventricle. The latter expels it to the great *aorta*, the wide arterial trunk that is the commencement of the general systemic circulation. Although the aorta arises on the left and the pumonary artery on the right, this relation appears reversed from in front because the two vessels are intertwined.

INTERIOR OF THE HEART

The openings between the chambers, and between the ventricles and their great arteries, are guarded by a system of *valves*. Between each atrium and its ventricle is a parachute-like valve, the strands of which are attached to the ventricular wall, the flaps lying on the orifice between the chambers. On the right this is the *tricuspid valve*, with three flaps, on the left the *mitral* (*bicuspid*) with two. When the atria contract, the blood flow easily separates the flaps as it passes into the ventricles; but when the latter contract in *systole*, the flaps are ballooned out, the chordia tendons that hold the valves in place are tautened, and no reflux is possible, all the blood being directed to the arterial exits. The latter, the openings into the pulmonary artery and aorta from the right and left ventricles, are guarded by simpler valves consisting of three semilunar pockets facing upwards. These are easily pushed apart, but fall together when the ventricles relax in *diastole*, the weight of the blood column in each pocket forcing them into contact and blocking any reflux.

Disease may render any valve incompetent and, by overworking the heart muscle in an effort to compensate for reflux, lead to heart failure, particularly the aortic valve which, fortunately, can be relatively easily replaced by a synthetic or animal substitute.

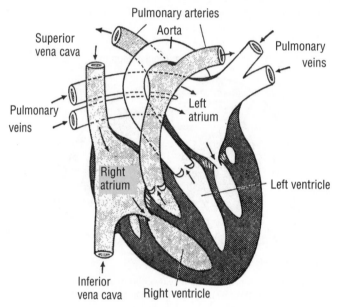

Figure 9.10 Diagram showing the interior of the heart, entry and exit of the great vessels, and the direction of blood flow

The diagram is artificial in that no section normally exposes all four chambers in this way

Great vessels of the thorax

The *superior vena cava* is formed by the junction of the *right* and *left brachiocephalic* veins, each bringing venous blood from one arm and one side of the head and neck. The right brachiocephalic and the vena cava run vertically, continuous with the right atrium. But the left brachiocephalic crosses from left to right in front of the trachea and above the aortic arch.

The *inferior vena cava* ascends from the abdomen through the central tendon of the diaphragm and has only a very short intrathoracic course before it enters the lower part of the right atrium.

The *pulmonary artery* arises at the upper border of the heart from the summit of the left ventricle and twines round the aorta. Behind the arch it divides into right and left branches,

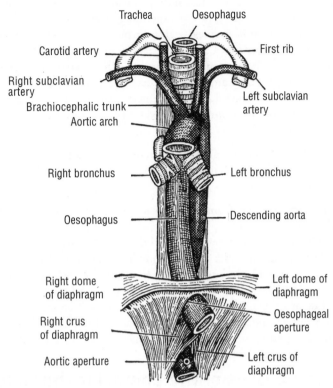

Figure 9.11 Structures of the posterior thoracic wall: oesophagus and descending aorta
The heart and lungs have been removed (after *Gray*)

which enter the roots of the lungs with the pulmonary veins and bronchi. The *pulmonary veins* are not seen from in front; two on each side, they drain into the back of the left atrium.

AORTA

This is the largest artery of the body, the beginning of the general or *systemic* circulation – as opposed to the local *pulmonary* circulation. It arises at the upper border of the heart from the apex of the left ventricle and its subsequent course is as follows:

1 a short *ascending portion* running up to the left of the superior vena cava

2 the *aortic arch*, curving horizontally backward in front of the lower part of the trachea and left bronchus

3 the *descending aorta*, running down on the vertebral column at the back of the thoracic cavity to traverse the diaphragm, where it continues as the abdominal aorta.

▶ Branches of the aorta and corresponding veins

From the summit of the aortic arch three great vessels arise. On the right is the *brachiocephalic trunk*, which divides rapidly into the *right subclavian artery* for the arm and the right common carotid for the head and neck. On the left there is no corresponding trunk; the carotid and subclavian spring directly from the arch. Only the lower parts of these vessels lie within the thorax, the level of the brachiocephalic bifurcation being at the inner end of the right clavicle; they are closely applied to the trachea. The corresponding veins lie more superficially and there is a *brachiocephalic vein* on both sides; each is formed by the subclavian vein and the vessel equivalent to the common carotid artery, the *internal jugular vein*. The small but vital *coronary arteries* arise at the very beginning of the aorta, immediately above the semilunar valves. There is a right and left vessel and any obstruction impairs the circulation in the muscle of the heart wall, giving rise to angina during exertion or a myocardial infarction of variable size, sometimes fatal. The coronaries can be replaced by grafts from the superficial veins of the upper or lower limb.

The *descending aorta* gives off on each side the *intercostal arteries* that encircle the chest in the intercostal spaces; the corresponding veins do not enter the venae cavae directly, but through intervening channels, the *azygos veins*.

The aortic arch forms a great sweep from front to back of the chest, with an upward convexity from which arise the main arteries of the head, neck and upper limbs. The concavity of the arch embraces the pulmonary arteries and veins and the bronchi, entering the lung roots. The arch inclines to the left, so that the descending portion is to the left of the midline and nearer the left lung. There is also a little artery running the length of the back of the sternum from above down on both sides. This is the *internal mammary artery*, a branch of the subclavian in the neck. It ends below by entering the sheath of the rectus muscle as the *superior epigastric artery* to anastomose with the *inferior epigastric* from below. The internal mammary artery can sometimes be used to join up with the healthy segment of an otherwise diseased coronary.

Remaining structures of the posterior thoracic wall

The *oesophagus* or gullet enters the chest from the neck at the thoracic inlet, lying immediately in front of the vertebral column. It travels down the back of the cavity between the roots of the lungs to traverse the diaphragm. It inclines forward and to the left in this passage, allowing the descending aorta to intervene between its lower portion and the spine; and it is more closely related to the right than the left lung. It is a hollow muscular tube down which food is propelled by waves of contraction or *peristalsis*; unlike most other parts of the bowel, it has no serous coat whatever as it lies entirely away from the pleura at the back of the mediasitinum.

The *thoracic duct* begins below as the upward continuation through the aortic aperture of the diaphragm of the cisterna chyli of the abdomen (Figure 5.13), the channel receiving digested fat from the bowel. The duct ascends on the right of the vertebral column behind the oesophagus, collecting lymphatics from the lungs; it crosses to the left side at the level of the aortic arch, enters the root of the neck on the left side, and discharges into the junction of the left subclavian and internal jugular veins.

Nerves within the thorax

There are four important nerve trunks in the thorax.

1 The *vagus nerve* on each side. This is the tenth cranial nerve, which has travelled down the neck via the thoracic inlet. The right *vagus* runs alongside the trachea, passes behind the right lung hilum, where it breaks up into a *pulmonary plexus*, reforms below as a single trunk and accompanies the oesophagus through the diaphragm. The *left vagus* runs with the left carotid and subclavian arteries, crosses the aortic arch, passes behind the lung root to form a similar plexus, reforms and completes its thoracic course beside the oesophagus.

The vagi are part of the *parasympathetic* component of the *autonomic nervous system* described in Chapter 20, concerned with automatic regulation of the viscera. In the chest they give branches to heart, lungs and oesophagus.

2 The *sympathetic trunks*, one on each side, lie 3–5 cm (1/2–2 inches) away from the vertebral bodies. They consist of a chain of *ganglia* (relay stations of nerve cells and fibres) lying on the necks of the ribs behind the parietal pleura and connected by longitudinal fibres. This chain is continuous with that in the neck above and leaves the thorax below to become the abdominal sympathetic. It too is part of the autonomic nervous system and shares with the vagus the dual control of the viscera. Several branches run down from the main trunk. These are the *splanchnic nerves*, which pierce the diaphragm to join the *coeliac plexus* surrounding the beginning of the abdominal aorta. (Figure 9.8)

3 The *intercostal nerves*, derived from the spinal cord, run round the chest wall in the intercostal spaces. At their origin they lie deeply behind the pleura and are connected to the sympathetic ganglia by communicating twigs.

4 The *phrenic nerves* arise in the lower part of the neck to supply the diaphragm. The *right phrenic* runs down alongside the superior vena cava and right side of the heart; the *left* crosses the aortic arch and left side of the heart. Both pierce the diaphragm and break up on its undersurface.

Test yourself

1 How many true ribs are there in the body?
   a 10
   b 12
   c 14
   d 16

2 What happens to the shape of the diaphragm during inspiration?
   a It flattens
   b It domes upwards
   c It domes downwards
   d It gets smaller

3 Which muscle aids the deltoid in shoulder abduction?
   a Pectoralis major
   b Pectoralis minor
   c Trapezius
   d Latissimus dorsi

4 Which membrane encloses the heart?
   a Mediastinum
   b Pericardium
   c Peritoneum
   d Pareital layer

5 Which valve lies between the right atria and the right ventricle?
   a Bicuspid
   b Mitral
   c Tricuspid
   d Semilunar

6 Identify the blood vessels that supply the heart with blood.
   a Coronary arteries
   b Subclavian artery
   c Brachiocephalic trunk
   d Azygos veins

7 Which arteries arise from the descending aorta?
   a Mammary
   b Intercostals
   c Inferior epigastric
   d Superior epigastric

8 Identify the location of the aortic arch.
   a Left of the superior vena cava
   b Runs down the vertebral column
   c Curves backwards in front of the left bronchus
   d Curves forwards behind the trachea

9 Identify a blood vessel that can be used as grafts to replace coronary arteries.
   a Azygos veins
   b Superficial vein of the lower limb
   c Descending aorta
   d Internal jugular vein

10 Which thoracic nerves are part of the parasympathetic nervous system?
   a Splanchnic nerves
   b Intercostal nerves
   c Left vagus nerve
   d Phrenic nerves

10

Regional anatomy: head and neck

In this chapter you will learn about:

▶ *the head, including the skull and base of the skull*

▶ *the surface anatomy of the head*

▶ *the face, facial vessels and nerves*

▶ *the mouth and the teeth*

▶ *the neck and its surface anatomy*

▶ *the pharynx, larynx and trachea*

▶ *the great vessels of the neck*

▶ *the nerves of the neck*

▶ *the lymph nodes of the head and neck.*

The head

SKULL

The skull is a jigsaw of bones, fitting together at the immobile fibrous *sutures* which are obliterated by ossification between the ages of 30 and 40. Although complicated, it is best understood by reducing it to three essential parts: cranium, facial skeleton and mandible.

The *cranium* is the box enclosing the brain and its membranes. It has a domed *vault* and a *base* set deep beneath the soft structures.

The *facial skeleton* is attached to the underside of the base anteriorly and includes the nasal bone, maxilla (upper jaw) and others.

The *mandible*, or lower jaw, is entirely separate and is slung to the underside of the skull base at the back.

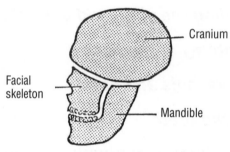

Figure 10.1 The basic plan of the skull

The skull bones are mostly flat; in several places, particularly the vault, they have an *outer* and *inner table* of compact bone sandwiching a red-marrow-filled layer – the *diploë*. At certain sites they are expanded by internal air-spaces or *sinuses*, or honeycombed by *air-cells*.

The vault is made up of four main bones:

▶ the frontal

▶ two parietals

▶ the single occipital.

The *frontal* is the substance of the forehead and overhangs the orbital cavities in front. The *parietals* lie on each side above the temples and the *occipital* at the back. The *coronal suture* separates frontal and parietals. The *sagittal suture* runs from front to back between the parietals in the midline. The *lambdoid suture* is the three-way junction of parietals and occipital. At birth, the angles of junction at either end of the sagittal suture are enclosed by bone and filled in with soft membrane, the *anterior* and *posterior fontanelles*. The posterior closes by the end of the sixth month, the anterior not until the second year.

Figures 10.3 and 10.4 are anterior and lateral views of the whole skull. Note in the *anterior view* the frontal bone coming down as a roof over the *orbital cavities*, the bony eye sockets; the openings of the *nasal cavities*, with the *nasal bones* above and the bony *nasal septum* separating the two sides; and the *maxilla*, carrying the upper teeth. Note in the *lateral view*, the side aspect of the vault and how the *temporal bone*, with the *parietal*, helps to form the

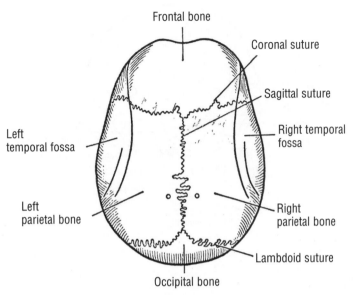

Figure 10.2 The vault of the skull, or vertex, from above

lower part of the side-wall in the hollow above the cheek; the cheek bone or *zygoma*, arching across from maxilla to temporal; the *mastoid process* of the temporal behind the bony external opening (*external auditory meatus*) of the ear canal; the occipital bone behind; and the *styloid process* of the mastoid jutting out from the base below.

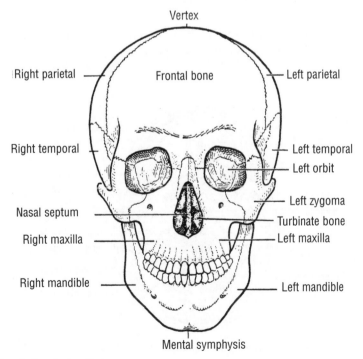

Figure 10.3 The skull, anterior view

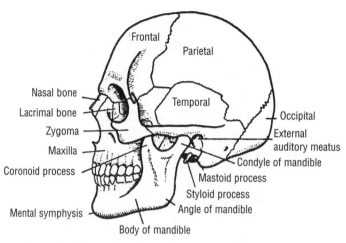

Figure 10.4 Lateral view of skull

THE SKULL BASE

Figure 10.5 shows the upper surface of the skull base after the skull cap has been removed. Note the great aperture, the *foramen magnum* nearer the back, where the spinal cord is continuous with the medulla of the hindbrain; and the numerous smaller *foramina*, where veins and cranial nerves exit and the cerebral arteries enter. The inner table is grooved by narrow channels for the arteries supplying the brain membranes, or by wider paths for the venous *sinuses* which run between the layers of the *dura mater*, the tough fibrous outer membrane covering the brain. The base is divided on each side into three pockets or *fossae*, which become deeper and more extensive from front to back.

The *anterior fossa*, in front, houses the frontal lobe of the brain. Its floor is the roof of the orbit and nose. Where the two sides of the anterior fossa meet in the midline, tiny perforations exist for passage of the olfactory nerves from nose to brain.

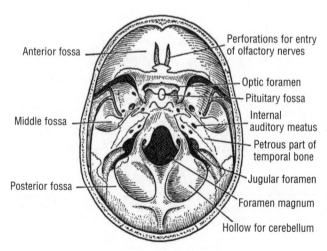

Figure 10.5 The skull base, seen from within, after removal of the skull cap

The *middle fossa* houses the temporal lobe of the brain; through a small opening just behind the anterior fossa, the optic nerve connects eyeball and brain. In its floor are large foramina for the exit of the fifth or trigeminal nerve, concerned with sensation in face and jaws and movement of the masticatory muscles, and for entry of the internal carotid artery into the skull. Between the two halves of the middle fossa a little socket embedded in a small platform is the *pituitary fossa* housing the pituitary gland.

The *posterior fossa* is separated from the middle by a massive oblique, bony ridge, the *petrous* portion of the temporal bone, enclosing the middle and inner ear cavities, and the passage of the seventh (facial) and eighth (auditory) nerves. The posterior fossa is mainly composed of the occipital bone; it contains the foramen magnum and a large foramen for the exit of the internal jugular vein, accompanied by various nerves.

On the undersurface of the base, on each side of the foramen magnum, the occiput carries an articular process or *condyle*, forming a joint with the atlas.

The *mandible* is formed by two halves joined in the midline at the chin, the *mental symphysis*. Each half consists of a horizontal *body* carrying the lower teeth on its *alveolar margin* (bony gum), and a vertical *ramus*. Body and ramus meet at the *angle*. At the upper end of the ramus are the pointed *coronoid process* and the rounded *condyle*, which articulates with the undersurface of the temporal bone at the *temporomandibular joint* in front of the ear (Figure 10.4).

The *air sinuses* of the skull are accessory extensions of the nasal cavities, and their lining mucous membrane is continuous though the connecting passages and may be intricate. The main ones are the *frontal* and *maxillary*; there are also certain groups deep in the skull – *ethmoids* and *sphenoids*. The maxilla is entirely hollowed out by a great cavity, the *maxillary antrum*. The *air-cells* which honeycomb the mastoid process are independent structures.

The *hyoid* is a little bone which can be felt in the neck in front, between chin and larynx, giving attachments to the muscles of the floor of the mouth and the tongue. It has a tiny body with spreading wings (Figure 10.9).

SURFACE ANATOMY OF THE HEAD
Most of the external surfaces of the cranium, facial skeleton and mandible are easily palpable; the base is inaccessible. In the *cranium* the vault is easily felt through the scalp, which moves over the bone. In front, the overhanging ridges of the frontal bone, the upper margins of the orbital cavities, beneath the eyebrows, meet in the midline at the root of the nose. There are four bulges on the vault: the frontal eminences of the forehead and the parietal eminences above the ears. The lowest accessible part of the skull in the midline posteriorly is the *external occipital protuberance*, which is reached by tracing upwards the median furrow between the muscle masses of the nape of the neck.

In the face, the *nasal bones* brace the bridge of the nose. The bulbous tip of the nose is supported only by cartilage. A finger in the nostril touches the *septum* between the two cavities; this is cartilage in front and the true bony septum is further back. The *zygomatic arch*, the cheek bone, is felt running between orbit and ear. Above is it the depressed *temporal fossa*, in which lies the *temporalis muscle*, which shuts the mouth and can be felt on clenching the teeth.

The *external ear* is the flap of skin and elastic fibrocartilage seen externally. Its aperture is the *external auditory meatus*, which is cartilaginous to begin with and bony deeper in. Behind the ear is the *mastoid process*, with its tip below; just in front of the ear, below the zygoma, the *condyle* of the *mandible* can be felt moving at the *temporomandibular joint* when the mouth is opened and closed. The joint can dislocate in yawning.

Body, ramus and *angle* of the mandible are all accessible, though the coronoid process is not palpable. The outer surface of the ramus is covered by the *masseter muscle*, which helps close the mouth and stands out on biting. If this muscle is contracted by clenching the teeth, a small tubular structure can be rolled under the skin parallel to and a finger's breadth below the zygoma, the duct of the *parotid salivary gland* running forward to open into the mouth.

The *scalp* includes all the structures overlying the vault. Its skin is profusely supplied with hair follicles and the subcutaneous fat is thin and dense. Beneath is a muscle sheet, most developed in the frontal and occipital regions as the *frontalis* and *occipitalis* bellies, connected by an aponeurotic sheet – the *galea* – stretching over the vault. The galea is separated from the bone by loose areolar tissue and all the layers of the scalp move as one over the skull when its muscles contract, as in raising the eyebrows. The scalp *veins* commmunicate freely with those of the diploë, the venous sinuses of the dura and those of the brain. The arterial supply is profuse.

FACE

The *facial skin* is thin, mobile and vascular and the muscles lie immediately beneath (Figure 10.6). Eyes and mouth are surrounded by circular muscles, the *orbicularis oculi* and *orbicularis oris*, which close the eyelids and lips. Small muscles attached to the nasal cartilages wrinkle the nose and dilate the nostrils. Others lift or depress the corners of the mouth, and the *platysma* of the neck is a broad sheet which spreads up over the mandible to share control of expression.

All these are supplied by the seventh (*facial*) nerve, which leaves the base of the skull in front of the mastoid, runs forward in the parotid gland and breaks up into branches to face and scalp at its anterior border.

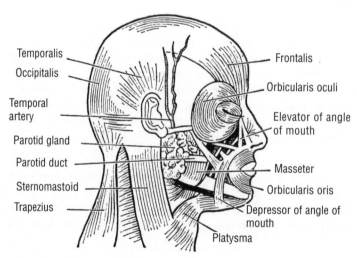

Figure 10.6 Facial muscles and associated structures

The muscles of *mastication* are a separate, deeper group, supplied by the fifth (*trigeminal*) cranial nerve. They include the *masseter* and *temporalis* – responsible for biting – and an *internal* and *external pterygoid* lying on the deep surface of the ramus, responsible for lateral and rotary chewing movements. A very deep muscle, the *buccinator*, runs transversely between mouth and masseter in the substance of the cheek.

The *parotid salivary gland* lies under the skin in front of the ear, covering the back of the masseter, and reaches up to the zygoma and down to the mandibular angle. The facial nerve is embedded in it, and it overlies the internal carotid artery and internal jugular vein as they travel between neck and skull base. Inflammation and swelling of the parotid occur in mumps. The parotid duct runs forward on the masseter to open into the mouth by piercing the buccinator opposite the second molar tooth.

The other external salivary gland is the smaller *submandibular*, which lies lower down in the uppermost part of the neck, partly under cover of the angle of the mandible; its duct pierces the floor of the mouth.

▶ **Facial vessels and nerves**

The *superficial temporal artery* runs in front of the ear, where it can be felt pulsating, to the scalp. The *facial artery* ascends from the neck over the body of the mandible, obliquely upward to the angles of eye and nose, giving branches to the lips and lids. Both vessels are branches of the external carotid.

The seventh (*facial*) cranial nerve is mainly motor to the muscles of expression.

The sensation of face and scalp is mediated by the fifth (*trigeminal*) nerve, which also supplies the masticatory muscles. The *lymph drainage* of the face is to groups of parotid and submandibular nodes, that from the lower lip goes to small nodes under the chin.

MOUTH (ORAL CAVITY)

This is the first part of the digestive tract. It is formed by the cheeks and lips at the sides, roofed by the palate, and its floor is the muscular *oral diaphragm*, separating it from the neck, in which the root of the tongue is embedded. The oral cavity is formed by a glandular mucosa and is continuous behind with the pharynx. The *vestibule* is the space between lips and teeth.

The red surface of the *lips* is the mucous membrane, sharply demarcated from the skin at the mucocutaneous junction. Each lip has a branch of the facial artery running parallel to its free margin and felt pulsating when the lip is gripped between finger and thumb. Each is attached at its midpoint to the gum by a fold of mucous membrane.

TEETH

These are set in the opposed (*alveolar*) margins of the upper and lower jaws, where mucosa and periosteum are firmly blended to form the gums. The first, temporary, set of milk teeth erupt at intervals throughout the first two years; the second, permanent, set begin to replace them at six years and are complete by 25 years, although the last molar or wisdom tooth is often delayed, impacted or may never appear.

The teeth are named, from front to back, as *incisors* (seizing), *canines* (tearing), *premolars* and *molars* (grinding) and are symmetric between either side and in upper and lower jaws, so that each quarter of the whole set is the same. This may be expressed as *dental formulae* for the permanent dentition:

	Molar	Premolar	Canine	Incisor	Incisor	Canine	Premolar	Molar	
Upper	3	2	1	2	2	1	2	3	
Lower	3	2	1	2	2	1	2	3	= 32

and the milk dentition:

	Molar	Premolar	Canine	Incisor	Incisor	Canine	Premolar	Molar	
Upper	2	0	1	2	2	1	0	2	
Lower	2	0	1	2	2	1	0	2	= 20

Note the absence of premolars in the latter.

Each tooth has a *crown*, projecting beyond the gum; a *root*, embedded in the aveolus; and an intermediate *neck*. Longitudinal section shows the central *pulp cavity*, outside this the *dentine* or ivory, the bulk of the tooth, and an outer coating of hard *enamel*. The vessels and nerves enter the pulp through a small foramen at the tip of the root (Figure 10.7).

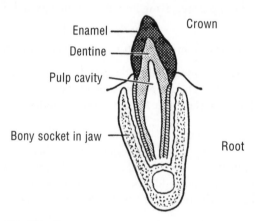

Figure 10.7 Longitudinal section of tooth

The *palate* is the roof of the mouth and floor of the nose. It consists of the bony *hard palate* in front and the mobile, muscular *soft palate* behind, projecting into the pharynx. The soft palate is arched, with a supporting pillar attached on either side of the back of the tongue; each pillar splits to enclose the *tonsil*, a mass of lymphoid tissue. From the free posterior edge of the soft palate there hangs down a small conical process, the *uvula*.

In the *floor of the mouth* a fold or *frenulum* is seen tethering the underside of the tongue to the mucous membrane; the duct of the submandibular salivary gland opens on each side at a little papilla. Between the frenulum, and the jaw on each side, a loose mucosal flap can be seen, or rolled by the tongue; this is the sublingual fold overlying the sublingual salivary gland, whose small ducts open directly into the mouth. Parotid, submaxillary and sublingual glands make up the three main salivary glands but many tiny mucous glands are scattered around.

The *tongue* is a muscular organ concerned with swallowing, taste and speech. Its anterior portion lies horizontally in the floor of the mouth; the back curves round vertically to form the anterior wall of part of the pharynx. The root of the tongue is embedded in the oral diaphragm; it is attached by various muscles to the hyoid bone and mandible, and to the soft palate by the pillars of the latter.

The *dorsum* is the upper surface, with a number of *papillae*, hair-like processes, and *taste buds*. The tip rests against the back of the upper teeth and the undersurface is attached to the floor of the mouth by the frenulum. The organ is mainly muscle, with right and left halves separated by a fibrous partition; but it has a mucosal covering specialized for the sensation of taste.

The neck

The neck connects the head and trunk. Its bony framework is the seven cervical vertebrae behind, and it is traversed by the food and air passages on their way to the thorax and the great vessels and nerves running between the thoracic inlet and the base of the skull.

At the root of the neck on each side is an outflow of nerves and vessels over the first rib to the upper limb.

Cross-section of the neck shows the relative arrangement of these structures. But whereas a section above the level of the sixth cervical vertebra shows the air and food passages as larynx and pharynx, at a lower level these have become the trachea and oesophagus.

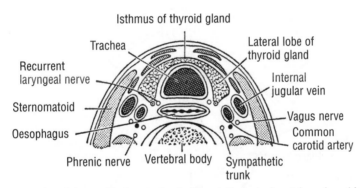

Figure 10.8 Cross-section of the neck below the level of the sixth cervical vertebra, viewed from above
The section does not include the back half of the vertebra and associated spinal muscles (after *Gray*)

The main points are (Figure 10.8):

1 the supporting *cervical vertebrae*, nearer the back

2 the thick *muscle masses* applied to the back of the spine

3 the *oesophagus*, immediately in front of the spine

4 the *trachea* between skin and oesophagus

5 the compartments of deep fascia on either side, the *carotid sheaths* in which run the *carotid artery*, *internal jugular vein* and *vagus nerve*.

Longitudinal section indicates the vertical arrangements and the transition from larynx to trachea and from pharynx to oesophagus (Figure 10.10).

TRIANGLES

The neck is divided into several triangles by various muscles. The key is the *sternomastoid*, a long flat muscle on each side whose origin is by two heads: from the inner end of the clavicle and upper border of the manubrium. It passes up and out to its insertion at the mastoid. The area in front of the muscle is the anterior triangle and that behind the posterior triangle (Figure 10.9).

The *anterior triangle* is bounded by the anterior midline of the neck and the anterior border of the sternomastoid, meeting at its apex below; its base, above, is the body of the mandible. It is subdivided into four smaller triangles.

▶ The *muscular triangle* contains several small muscles which depress the larynx in swallowing.

▶ The *carotid triangle* contains the upper part of the common carotid artery as it divides into internal and external carotids, the internal jugular vein and important nerves.

▶ The *digastric triangle* contains the submandibular gland.

▶ The *submental triangle* contains a few small lymph nodes.

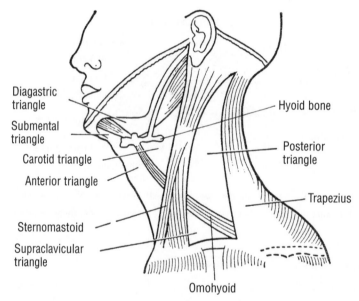

Diagastric triangle

Submental triangle

Carotid triangle

Anterior triangle

Sternomastoid

Supraclavicular triangle

Hyoid bone

Posterior triangle

Trapezius

Omohyoid

Figure 10.9 Muscular triangles of the neck (after *Gray*)

The *posterior triangle* is bounded by the back of the sternomastoid in front and the front of the trapezius behind, meeting at the apex at the occipital bone; its base is the clavicle. It is subdivided by the tiny omohyoid muscle into an upper *occipital* and a lower *supraclavicular* triangle. The former is mainly muscular, but the latter contains the subclavian vessels and brachial plexus of nerves on their way to the arm.

SURFACE ANATOMY OF THE NECK

▶ Anterior

The *sternomastoid* is very prominent, especially when tautened by turning the head to the opposite side, when its sternal tendon can be felt between finger and thumb. Between the sternal heads of the two muscles is the *suprasternal notch*, a hollow at the upper border of the manubrium. The *external jugular vein* can be seen running obliquely down and back under the skin, superficial to the sternomastoid.

The main structures in the midline of the front of the neck are as follows.

▶ In the hollow under the chin is the small *hyoid bone*, felt when the throat is grasped, and sometimes fractured in manual strangulation.

▶ Immediately below is the *laryngeal prominence*, or Adam's apple, composed of the two halves of the *thyroid cartilage* meeting at a V-angle in front, with a notch at their upper border.

▶ Below this is the *cricoid cartilage* of the larynx, at the level of the sixth cervical vertebra.

▶ Lower still are the uppermost cartilaginous rings of the *trachea* (see Figure 10.11).

The pulsation of the carotid arteries is felt by pressure along the anterior borders of the sternomastoids. Behind this muscle, the posterior triangle is marked by a depression of the skin, and at the base of this, behind the upper border of the clavicle, the *subclavian artery* can be felt beating.

▶ Posterior

At the back of the neck, the median *longitudinal furrow* between the muscle bellies of the uppermost part of the erector spinae muscles overlies the *spinous processes* of the cervical vertebrae.

FASCIAE AND MUSCLES OF THE NECK

The deep fascia in the neck forms a continuous encircling sheet. There are also deeper concentric layers around the trachea and, most deeply, the vertebral bodies. The muscles are partitioned by fascia, and a condensation of the membrane on each side forms the *carotid sheath* containing the common carotid artery, internal jugular vein and vagus nerve.

The cervical muscles are grouped as follows.

1 *Superficial: platysma, trapezius, sternomastoid.* The trapezius and sternomastoid are both supplied by the eleventh (*accessory*) cranial nerve and have already been discussed. Each sternomastoid acts by bringing the ear down to the shoulder and turning the chin to the other side; both acting together flex the cervical spine, bringing the chin down to the sternum.

The *platysma* is a muscle of facial expression. A broad sheet without bony attachment, it arises from below the clavicle and spreads up the side of the neck and over the mandible to blend with the facial muscles at the angle of the mouth.

2 The *suprahyoid* muscles are a small group between hyoid and mandible, forming the floor of the mouth and supporting the base of the tongue.

3 The *infrahyoid* group are strap-like muscles, arising from the back of the manubrium and ascending to the hyoid bone and thyroid cartilage on each side of the midline. They depress the larynx in swallowing.

4 The *prevertebral* muscles are the deepest, running in front of the vertebral bodies behind the other structures of the neck; they help flex the cervical spine.

5 The *scalene* muscles are found in the posterior triangle. They run down and out from the transverse processes of the cervical vertebrae to the first and second ribs. They elevate these ribs in inspiration and are close to the subclavian vessels and brachial plexus as these run out over the first rib.

PHARYNX

In the lower part of the neck and in the chest the air passages (larynx and trachea) and food passage (oesophagus) are separate. But in the upper part of the neck there is a continuous cavity extending up to the base of the skull. This is the *pharynx*, situated immediately in front of the vertebral column (Figure 5.9).

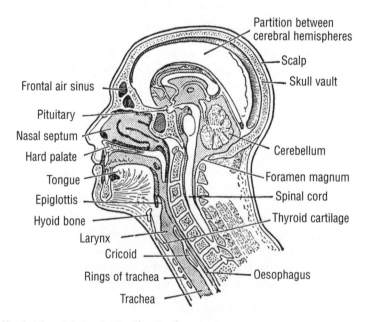

Figure 10.10 Head and neck in longitudinal (sagittal) section

It lies successively behind the nasal cavity, the back of the mouth and the opening of the larynx, and is therefore divided into *nasopharynx*, *oral pharynx* and *laryngopharynx*. Its upper part reaches the skull base and below it is continuous with the oesophagus at the level of the sixth cervical vertebra.

The pharyngeal wall is a muscular cylinder designed to propel food between mouth and oesophagus, and consists of three *constrictor* muscles – *superior*, *middle* and *inferior* – corresponding to the divisions of the cavity. In swallowing it is essential to prevent food passing up the nose round the back edge of the soft palate or down into the larynx. Therefore, the soft palate is opposed to the posterior pharyngeal wall to shut off the nasopharynx; and the *epiglottis*, a flap of elastic fibrocartilage situated at the summit of the larynx behind the root of the tongue is bent over backwards to close off the entry into the larynx. During the act, the larynx moves upwards and then descends again (Figure 14.1).

LARYNX

The larynx is the organ of voice and part of the air passages. It opens above into the pharynx at the back of the tongue and is continuous with the trachea below. It projects forward in the upper part of the neck under the skin.

The organ is a framework of cartilages connected by ligaments and membranes to form a resonating chamber containing the vocal cords. The *thyroid* cartilage is the largest; its two halves meet at the midline as the Adam's apple, with a superior notch. From the back of this notch arises the stalk of the *epiglottis*, a leaf-like fibrocartilage plate sticking up behind the hyoid and the base of the tongue. The *cricoid* cartilage at the lower end is a signet ring-shaped structure and immediately below is the first ring of the trachea (see Figure 10.11).

The *vocal cords* are a pair of mucosal folds, strengthened by an underlying ligament, one on the deep surface of each thyroid leaf, stretching from front to back. The muscles associated with the larynx are *extrinsic*, e.g. the infrahyoid group, designed to move the organ as a whole as in swallowing, or *intrinsic*, concerned with regulating tension in the vocal cords. The latter may lie relaxed at the sides of the cavity, meet at the midline and shut off the airway, or occupy an intermediate position, giving gradations of pitch. The interval between them is the *glottis*, the narrowest point of the respiratory tract.

TRACHEA

The windpipe is a hollow cartilaginous and membranous tube, some 10 cm (4 inches) long, extending from the larynx through the thoracic inlet to its bifurcation into the bronchi. It is ⌒-shaped, with the convexity under the skin of the front of the neck and the base resting on the oesophagus, which separates it from the vertebral column. It consists of a number of cartilaginous rings, incomplete behind, with connecting membranes. In the neck the second, third and fourth rings are crossed by the isthmus connecting the two lobes of the thyroid gland. The common carotid arteries lie beside the trachea.

It has a lining mucous membrane which is ciliated, i.e. each cell bears a hair-like process that wafts foreign irritants and mucus up towards the base of the tongue, where they are swallowed.

The oesophagus is a muscular channel about 25 cm (10 inches) long for the transfer of food from the pharynx to the stomach. It begins at the level of the cricoid cartilage as the continuation of the lowest portion of the pharynx. It descends in front of the vertebral column behind the trachea and enters the thoracic inlet.

The thyroid gland is one of the group of hormone-secreting endocrine glands. It lies in the lower part of the neck as a right and left lobe applied to either side of the trachea, with a connecting *isthmus*. Each lobe is pyramidal, apex upward, at the sides of the cricoid and thyroid cartilages; laterally it overlaps the carotid sheath. The gland is composed of numerous tiny vesicles containing *thyroxin*, the secretion essential for growth and activity; deficiency leads to *cretinism* in children and retardation (*myxoedema*) in adults. It also secretes the thyrocalcitonin necessary for calcification of bone. Any enlargement of the organ is called a *goitre*.

The parathyroid glands are four pea-like bodies embedded in the lobes of the thyroid, an upper and a lower on each side (see Figure 21.1). They are endocrine glands controlling the utilization of calcium, and removal or deficiency causes the convulsive syndrome known as *tetany*.

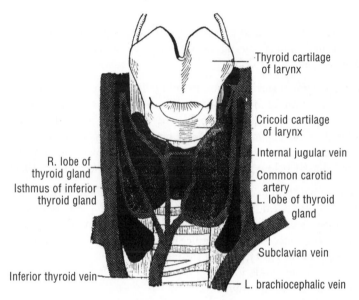

Figure 10.11 Larynx, thyroid gland, great vessels of the neck, viewed from above (after *Bruce and Walmsley*)

GREAT VESSELS OF THE NECK

On the *right* side the *brachiocephalic trunk* from the aorta forks at the level of the sternoclavicular joint into subclavian and common carotid arteries; on the *left* there is no brachiocephalic trunk, the subclavian and carotid arising directly from the aorta.

In the neck the *common carotid* ascends on either side of the trachea and thyroid gland to the upper border of the thyroid cartilage, where it divides into internal and external branches. It runs in the carotid sheath, with the internal jugular vein lateral to it and with the tenth cranial (*vagus*) nerve. Just below the bifurcation is a dilatation, the *carotid sinus*, where the vessel walls are sensitive to changes in blood pressure; this sinus maintains reflex control of the circulation to the brain. Apart from its terminal division, the common carotid has no branches.

The *internal carotid* continues to the base of the skull, enters the cranium and supplies the brain and the contents of the orbit.

The *external carotid* supplies the neck and surface of face and head. It gives large branches to the thyroid, larynx and tongue, also the *facial* artery, and ends by running up behind the ramus of the mandible in the substance of the parotid gland to end as the *superficial temporal* artery to the scalp. This can be felt pulsating just in front of the ear over the zygoma. It also supplies the main arteries to the jaws.

The *internal jugular vein* corresponds to the common and internal portions of the carotid artery. It leaves the base of the skull as a continuation of the venous sinuses of the brain through a special foramen, in company with the ninth, tenth, eleventh and twelfth cranial nerves, and runs down the neck lateral to the carotid and under cover of the sternomastoid. It ends by joining with the subclavian vein behind the sternoclavicular joint to form the *brachiocephalic*, one on each side. There is no vein corresponding to the external carotid artery, but some veins from the face and scalp form the *external jugular vein*, which runs superficially over the sternomastoid under the skin and pierces the deep fascia above the clavicle to enter the subclavian vein.

NERVES OF THE NECK

These fall into the three groups.

▶ Cranial nerves

The ninth (*glossopharyngeal*), tenth (*vagus*), eleventh (*accessory*) and twelfth (*hypoglossal*) pairs of cranial nerves enter the neck from the cranium through foramina in the base of the skull. The first three travel with the internal jugular vein at this site.

The *glossopharyngeal nerve* is concerned with sensation and taste in the pharynx and back of the tongue and pierces the wall of the pharynx to supply its mucous membrane.

The *vagus nerve* runs down in the carotid sheath between the vein and artery, giving branches to pharynx and larynx. Its role in the chest and abdomen and its place in the autonomic system are referred to in Chapter 9. An important branch to the intrinsic muscles of the larynx is known as the *recurrent laryngeal nerve*, for it hooks round the underside of the subclavian artery on the right to turn back to the larynx. The right recurrent nerve is entirely within the neck, but the left descends to the thorax to hook round the aortic arch before ascending. Each recurrent nerve is intimately related to the back of the thyroid gland and may be damaged during operations here.

The *accessory nerve* supplies the sternomastoid and then runs down and back in the posterior triangle to reach the trapezius.

The *hypoglossal nerve* supplies the muscles of the tongue and floor of the mouth.

▶ Cervical spinal nerves

The roots of the spinal nerves emerge between the vertebrae on each side, a pair at every level. There are eight pairs, as the first nerve emerges between the skull and the atlas and the eighth below the seventh cervical vertebra. The upper roots form a small *cervical plexus* supplying various muscles and the phrenic nerve to the diaphragm. The lower roots, cervical pairs 5, 6, 7 and 8, together with the first thoracic root, form an important network, the *brachial plexus*, which lies in the lower part of the neck in the posterior triangle and gives off the nerves of the upper limb. This plexus forms a bundle with the subclavian artery and vein that passes over the first rib to the arm. Thus the plexus is partly in the neck above the clavicle and partly in the axilla.

Violent traction on the limb or sideways wrenching of the head, as may occur in traffic accidents, can stretch or rupture the nerves of the plexus or even avulse its roots from the spinal cord, leaving a limb paralysed and insensitive to a varying degree.

▶ Sympathetic chain

In the neck the *sympathetic nerve chain* consists of three *ganglia* – the *superior*, *middle* and *inferior* – with a connecting *sympathetic trunk*; these lie deeply on the prevertebral muscles on each side, behind the carotid sheath. The trunk is continuous with the thoracic portion of the sympathetic chain.

Lymph nodes of head and neck

The main collecting nodes for the head and face are: those associated with the ear, the pre- and postauricular groups; those on or in the parotid and submandibular glands; the submental nodes under the chin; the occipital nodes at the skull base behind; and deep nodes between pharynx and vertebral column.

There are superficial and deep chains of nodes and lymphatic vessels. The *superficial* nodes are grouped around the external jugular vein and accessory nerve in the posterior triangle; the *deep* nodes surround the internal jugular vein and carotid artery.

At the base of the neck, on the *right*, a major lymph trunk is formed from the vessels draining the right upper limb and right side of head and face; this discharges into the junction of the subclavian and internal jugular veins. On the *left*, the *thoracic duct*, with all the lymph from both lower limbs and from the abdominal and thoracic viscera, arrives in the posterior triangle, adds to itself the vessels of the left upper limb and left side of head and neck, and empties into the corresponding venous angle. The thoracic duct is much larger and more important than the right lymph trunk, but even when distended it is no thicker than a matchstick.

Test yourself

1 Which of the following bones is located in the vault of the skull?
    a Mandible
    b Cranium
    c Occipital
    d Maxilla

2 Identify the fossa which houses the temporal lobe of the brain.
    a Anterior
    b Middle
    c Posterior
    d Lateral

3 Which area of the skull forms a joint with the atlas?
    a Posterior fossa
    b Petrous
    c Occiput
    d Foramen magnum

4 What is the joint between the two halves of the mandible called?
    a Coronid process
    b Mental symphysis
    c Alveolar margin
    d Vertical ramus

5 Which joint can be dislocated when yawning?
    a Temporalis
    b Mastoid
    c Zygoma
    d Temporomandibular

6 What is the name of the muscle of the cheek?
    a Masseter
    b Interior pterygoid
    c Exterior pterygoid
    d Buccinator

7 Which of the following is *not* a salivary gland?
    a Parotid
    b Submaxillary
    c Sublingual
    d Thyroid

8 Which of the following triangles in the neck contains small lymph nodes?
    a Carotid
    b Digastric
    c Muscular
    d Anterior

9 Which of the following muscles in the neck help to elevate the rib during inspiration?

　a　Scalene

　b　Prevertebral

　c　Infrahyoid

　d　Suprahyoid

10 Which area is the narrowest point of the respiratory tract?

　a　Glottis

　b　Epiglottis

　c　Nasopharnyx

　d　Laryngopharynx

Part 2:
Physiology

11

General considerations

In this chapter you will learn about the history
of physiology.

The study of the structure of body systems can only be artificially separated from that of their behaviour during life, i.e. physiology. In this way it is complementary to anatomy, which is more concerned with shape and structure. Anatomy is a descriptive discipline, whereas physiology is experimental. We have to describe what the structures appear to do. But form and function influence each other reciprocally; a clear distinction would be both impossible and meaningless.

Observation and experiment may be performed on humans and other animals. Without animal experiment, modern physiology – and for that matter, modern medicine – could never have evolved. Fortunately, the basic behaviour of living matter is the same in microscopic unicellular creatures as in the cells of the human body. There is much more universality of physiologic process than there is of structure.

The human body carries out its functions in a constantly changing *external environment*. This is possible only if the *internal environment* – e.g. the composition of the body fluids, the core temperature – is kept constant despite external changes. This regulation of the inner state is of immense importance in physiology and is known as *homeostasis* (see Chapter 12). It can be achieved because the physical and chemical processes of life are largely reversible. The body maintains its internal equilibrium like a gyroscope, until it runs down into the irreversible changes of death.

The history of physiology

Physiology began with the ancient Greeks. Their observations were often surprisingly accurate, but they were misled by speculation, e.g. on the location of the 'vital humours' and the seats of the emotions and intelligence. Thus, *Aristotle*, in the fourth century BC, regarded the heart as the organ of reason, but made many accurate records of the development of animal embryos. The variations of the pulse were remarked by the physicians of his time, and the differences between arteries and veins, yet they had no concept of the circulation of the blood and their speculations about disease were little removed from magic.

After Aristotle, the Greek tradition was continued at Alexandria, particularly by *Eristratus*, who recognized the role of muscles in locomotion and the association of intelligence with the intricate convolutions of the brain. But he produced a confusing theory, based on the realities of respiration, of the permeation of the blood vessels and nerves by an essential spirit, or *pneuma*, diffused to the organs by the action of the heart.

In the sixteenth century, *Vesalius* rejected the teachings of the ancients and stressed the importance of observation and experiment He recognized that muscles were stimulated via their nerves, that they had antagonists which acted reciprocally, and that arterial pulsation was derived from the heartbeat. He also watched the movements of the lungs in dogs after removing parts of their ribs. Before Vesalius, the nature of the circulation had been obscured by *Galen's* insistence that blood passed form one side of the heart to the other, through invisible apertures in the intervening septum, but Vesalius asserted that the septum was impermeable. *Servetus* showed that transfer of the blood was effected through the pulmonary vessels and their ramifications in the lungs, leading to the discovery of the true nature of the circulation by *William Harvey*, a physician at St Bartholomew's Hospital, London, published in his *De Motu Cordis* in 1628. Harvey grasped the essentials, though he was as yet unable to see his postulated capillaries, the fine channels connecting arteries and veins. The evidence, though indirect, was conclusive: the output of the heart, which, at every beat, pumped about two ounces (56.7 grams) of blood into the aorta. The heart rate is some 72 beats to the minute, so over 241 kg of blood leaves the heart every hour, clearly impossible unless there *is* a circulation. This, with growth in chemical and physical knowledge and the arrival of the microscope, led to rapid advances. Harvey's capillaries were identified, the red corpuscles of the blood discovered, and blood transfusion in animals was performed in 1660, as Pepys noted.

The nature of the respiration was clarified by *Robert Boyle*, who showed that a part of the air, oxygen, was as essential to life as to combustion. Inspired air mixed with the blood in the lungs to give it its red arterial colour; deoxygenation in the rest of the body rendered it dark and venous. A country clergyman, *Stephen Hales*, one of a long line of distinguished amateur English eighteenth-century scientists, directly measured the blood pressure in horses, using glass tubes several feet high, laying the foundations of our knowledge of the haemodynamics of the body.

In the latter part of the eighteenth century the famous surgical brothers *William* and *John Hunter* made many observations on digestion, the coagulation of the blood, and the growth of bone. The discovery of vitamins was implicit in the Admiralty Order of 1795 for the issue of lime juice to check the ravages of scurvy in the navy.

Nugget

A slang name for British people was 'limeys' because they would take limes on board their ships to prevent scurvy.

In Italy at this time, *Galvani* investigated electrical phenomena in the body and showed that muscles could be made to contract by electrical stimulation.

In the nineteenth century we leave behind the slow advance by individual effort and co-opt the basic sciences. Recognition of the single cell as the essential unit of biologic structure and function; the discovery of diffusion and osmosis, crystalloids and colloids; the bridging of the gap between inorganic and organic by the synthesis of urea in 1823: all were of great significance. Further, the concept of *metabolism* was founded, i.e. the breakdown of foodstuffs by oxidation in the tissues to yield energy and carbon dioxide and other wastes, a process with as exact an equivalence of energy interchange as any reaction in a crucible, and no less subject to the laws of thermodynamics.

A picturesque episode is that of *Beaumont* in America, studying gastric function in Alexis St Martin, a trapper shot in the abdomen, with a permanent opening left between the stomach and the surface. This allowed direct observation of the reactions of the stomach to various foods and emotions, as shown by changes in secretion and motility.

It is to a Frenchman, *Claude Bernard*, that we owe the idea of the *constancy* of the internal environment, on which all life depends: a mechanism always acting to neutralize the effects of environmental changes, maintaining an equilibrium in which all the cells and fluids remain unaltered in their essential properties, a regulation achieved by the diffusion of water, oxygen, food and heat through the vascular system, the whole coordinated by the nervous system. In its regulation, the endocrine glands of internal secretion, such as the thyroid, that produce the hormones, were shown to have a cardinal role.

There has also been the development of our knowledge of the nervous system. *Sir Charles Sherrington*, at Oxford, demonstrated the self-regulation of posture and the activities pertaining to the different levels of the brain, that 'humming hive of dreams'. *Pavlov* discovered the conditioned reflexes linking mind and body, responses produced by reactions, not to the original stimuli, but to others learned to be associated with them. He suggested that human behaviour might even be considered as a complex of such reflexes. More recently the psychologists entered the field by showing the influence of the emotions on the most basic physiological activities, such as heartbeat and digestion.

The advances in the second half of the twentieth century are too numerous and too important to allow summary here. They include the *ultrastructure* of cells under the electron microscope and its relation to function, the technique of *positron emission tomography* which permits the study of metabolic processes and circulation, especially in the brain, after injection of radioactive isotopes, and the *genetic factors* disclosed by the unravelling of the human genome.

Test yourself

1 What name is given to the regulation of the inner state of the body?
   a Homeostasis
   b Vital humours
   c Dissection
   d Antagonist

2 Who recognized that muscles act reciprocally?
   a Aristotle
   b Eristratus
   c Vesalius
   d Galen

3 What essential spirit was thought to diffuse to the organs by the action of the heart?
   a Eristratus
   b Vital humours
   c Pneuma
   d Haemodynamics

4 Who showed that oxygen was important for respiration?
   a Harvey
   b Boyle
   c Pepys
   d Hales

5 What fruit was issued to the navy to prevent scurvy?
   a Apples
   b Limes
   c Pears
   d Oranges

12

Basic biophysics and biochemistry

In this chapter you will learn about:

▶ *homeostasis, and how the body responds to external changes*

▶ *the balance of intracellular and extracellular body fluids*

▶ *essential minerals*

▶ *nutrients*

▶ *enzymes*

▶ *metabolism.*

Homeostasis

The body appears to be a closed system but is exposed to a constantly changing *external environment*. These changes are buffered by responses that maintain a constant *internal environment*, which is known as *homeostasis*. As a result of this, the blood, lymph and tissue fluids and the cells they bathe and permeate are protected from change. All the vital mechanisms contribute to this, as in the following examples.

▶ Exercise produces heat and the body temperature tends to rise. This results in sweating to produce heat loss due to the evaporation of water, i.e. sweating is a physiologic cooling mechanism that operates whenever the temperature rises, whether from exertion or fever.

▶ Blood is normally slightly alkaline, with a pH of 7.4. During exertion the muscles produce carbon dioxide, CO_2, but there is little or no rise in the acidity of the blood. This is because of blood's chemical buffering properties, and also because the excess carbon dioxide is exhaled as it is formed. Hence the rapid breathing during exercise and the subsequent panting as the oxygen debt is paid off.

▶ Blood levels of sugar (glucose) remain constant during fasting, even though sugar is being burnt by the tissues and none is ingested. This is because of the formation of sugar from fat and protein in the body stores, mainly in the liver.

▶ Most of the water drunk enters the bloodstream, yet even large amounts of fluid do not lower the osmotic pressure of the blood by dilution. When excess water is consumed, it is sensed by special *osmoreceptors* at the base of the brain which affect the output of the pituitary hormone that controls the output of water by the kidney. Excretion of water in the urine thus keeps pace with absorption.

These examples illustrate how the internal environment is kept constant despite external stresses. Were these mechanisms to fail, as they sometimes do, the organism would be in danger, e.g. in high fever due to infections or when the blood sugar falls to a low level.

Body water

Water is the universal fluid medium of the body. It transports food, wastes and respiratory gases and allows them to diffuse in and out of cells. Water constitutes up to 84% of cell protoplasm and 60% of body weight. No substance can influence living cells until it is brought into watery solution.

The *total body water* is some 40 litres in a normal, healthy adult man, or two-thirds of body weight. This includes the water in free fluids, such as the blood plasma and the fluid of the tissue spaces, as well as that in the cells. The *extracellular water* is about 15 litres, one-fifth of body weight: three litres are contained in the plasma, 12 in the interstitial fluid. The remaining *intracellular water* amounts to 25 litres:

Total body water 40 l } Intracellular 25 l
Extracellular 15 l { Interstitial fluid 12 l
Plasma 3 l

The total content of extracellular fluid (ECF) varies a little, so that the body weight in the morning, after urination, fluctuates by up to a kilogram. The ECF is reduced by lack of water intake, or by increased output due to sweating, vomiting or diarrhoea. It increases in waterlogging of the tissues (*oedema*), as in starvation, heart failure or kidney failure.

There are major differences in composition of the ECF and the intracellular fluid (ICF). The *cations* of the ECF are mainly sodium, a little potassium, calcium and magnesium; those of the ICF are mainly potassium, some sodium and calcium, and rather more magnesium. The *anions* of the ECF are mainly chloride, with some bicarbonate; those of the ICF include some chloride, bicarbonate and sulphate, but are mainly organic acids, phosphate and proteins. Thus the chief difference is that sodium is the main cation outside the cells and potassium inside; and this difference is essential and constant. The ECF, or *internal environment*, is mainly a solution of NaCl. The bicarbonate of the ECF is controlled by respiration, which expels carbon dioxide and regulates the acidity of the fluid.

Body water is derived from the water content of solid food, the water drunk, and the metabolic water formed in the tissues by oxidation of the hydrogen in the food. Water is lost in the urine and faeces and by evaporation from the skin and lungs. The first may vary greatly, but the expired air is always saturated with water vapour.

The water balance in a sedentary adult male in a temperate climate is as follows:

▶ *Intake:* food 800 g, drink 1300 g, metabolic 250 g = 2350 g

▶ *Output:* urine 1500 g, faeces 25 g, evaporation 825 g = 2350 g

The evaporated water is important in cooling the body and accounts for a quarter of total heat loss. Because water loss from skin and lungs is obligatory, and because there is little or no reserve of water, a negative balance is easily produced if intake falls, in excess loss in diarrhoea and vomiting or sweating, or if loss increases from expiration in cold dry conditions or if kidney disease eliminates excess water. Death occurs in less than a week in total deprivation of water. However, it is impossible to create a positive water balance by drinking large amounts, for the excess water is lost in the urine.

From the *biophysical* angle, bodily function is concerned with the behaviour of molecules in aqueous solution. Thus, by diffusion, the salts of the blood or the sugar content of the cerebrospinal fluid are uniformly concentrated.

Dialysis occurs when certain membranes allow water and dissolved substances to pass freely. If such a membrane separates two solutions of different strengths, their concentrations are equalized by the transfer of water and solute in reverse directions.

Osmosis occurs when cell membranes are *semi-permeable*, allowing free passage to water but only to some solutes. If such a membrane separates two solutions of a substance that cannot traverse the membrane, water is drawn from the weaker to the stronger solution by osmosis. The *osmotic pressure* of a solution is the expression of the energy gradient set up by the tendency of water always to dialyse so as to dilute the stronger solution.

The behaviour of living cells is intimately affected by the relative osmotic pressures of the surrounding medium and of the cell substance, as the intervening cell membrane is semi-permeable. A red blood corpuscle, immersed in strong salt solution, shrivels as water is sucked out of it, but swells and bursts in a weaker solution or in water.

Isotonic solutions have an osmotic pressure for the surrounding medium which is identical to that of the cell protoplasm and as such leaves the cell unaffected. The stability of the cells is such that their osmotic pressures are identical and isotonic with a 0.9% solution of sodium chloride, or *normal saline*. Normal saline is thus identical osmotically to the blood and lymph and is used to bathe tissues during operations without harming them. It can also be given intravenously in an emergency to restore depleted blood volume, as in shock, though it will not remain in the circulation without colloid support.

Whatever the substance, solutions containing the same number of molecules per unit volume have the same osmotic pressure. Therefore, solutions of different substances of the same percentage strength have osmotic pressures inversely proportional to the sizes of their molecules, e.g. a 1% solution of protein, with its complex molecules, contains far fewer particles than a 1% solution of sugar, and its osmotic pressure is correspondingly less. Further, many substances called electrolytes *ionize* in solution, i.e. their molecules split or dissociate into two or more electrically charged *ions*. The osmotic pressure of such solutions is greater than would be expected from the size of the molecules, as it is proportional to the total number of *particles*.

Solutions have a further property of physiological importance due to their tendency to ionize. This property is their *chemical reaction*: acid, alkaline or neutral. Acidity is due to the presence of hydrogen ions (H) and alkalinity to hydroxyl ions (OH). The reaction of water is neutral because it produces these ions in equal amounts. In any solution the product of hydrogen and hydroxyl ions is constant, the two being inversely proportional, so that acidity or alkalinity of a solution may be estimated by measuring its hydrogen ion concentration. This is always a small figure (a litre of water contains only 10 g of such ions) and it is customary to use as an index the power of 10 as a positive number – the hydrogen ion exponent or **pH**.

Thus the pH of water is 7, and this is the centre point of chemical neutrality. The pH of a solution falls as its acidity rises. The pH of a solution increases as it becomes alkaline. A single step from, say, pH 7 to 8 indicates a tenfold increase of alkalinity. The reaction of the blood is normally just on the alkaline side, at pH 7.4. Cell functions are profoundly modifed by minor changes in chemical reaction. A small increase in acidity or alkalinity may bring the heart to a stop. Nevertheless, small amounts of acid and alkali are constantly being formed in life, and their effects are minimized by certain *buffer salts*, such as sodium bicarbonate, in the body fluids and protoplasm, which take them up automatically to prevent any change in reaction. The main buffer in the body fluids is *sodium phosphate*, and in the cells *sodium bicarbonate*; each can vary between its own acid and alkaline forms to compensate for any change in reaction of the medium. *Proteins* can function either as weak acids or as bases, and so stabilize the pH of the cells they compose.

Crystalloids and colloids

All substances fall into two classes:

▶ *crystalloids*, simple crystallizable compounds like sugar and salt which diffuse easily in water and traverse animal membranes

▶ *colloids*, complex materials like gelatin and egg white which crystallize with difficulty, if at all, diffuse slowly and cannot cross membranes.

The two can be separated by dialysis and their essential difference is the size of their molecules. The molecular weight of salt is 58.5, whereas that of proteins is of the order of 100,000.

Colloids may exist either as *sols* in apparent solution or as *gels*, a solid jelly state with water. Colloidal 'solutions' are really only suspensions of particles, a disperse phase in some fluid medium. These phases are reversible, as when milk, a suspension of oil in water, becomes butter, a suspension of water in oil.

A feature of colloidal 'solutions' is that their surface layer has a much greater concentration of colloid than the mass of the solution. This characterizes the bounding membranes of living cells, the membranes whose semi-permeability is responsible for osmotic transfers and for maintaining certain differences between the contained protoplasm and the surrounding medium.

Thus there is more potassium than sodium in red blood corpuscles, though the reverse is true of the plasma. Again, most cells have a slightly acid reaction of pH 6.8, though immersed in neutral or slightly alkaline media; these differences are maintained by the selective action of the cell membrane. But this selectivity is not purely physical, for the outer layer contains fatty materials and can be penetrated by agents soluble in fats, such as ether and alcohol.

Nugget

The fatty materials of the cell membrane of the brain allow anaesthetics to act on the brain.

Colloidal solutions are unstable and their particles may be precipitated by heat, changes in reaction, or the addition of salts. The *coagulation* of protein sols by heat is familiar in the albumin of egg white. A further important property of colloids is their power of *adsorption* of other substances due to the immense surface area of the dispersed colloid particles. Adsorption is the capacity to pick up and retain a substance without entering into chemical combination with it.

The basic physical factors that control the transference of substances across cell membranes are the following.

1 A difference of *hydrostatic pressure* between the two sides. Thus the osmotic pressure of the proteins dissolved in the blood plasma resists outward passage of water from the capillaries to the kidney tubules. This pressure is the equivalent of 30 mm of mercury. It is normally overcome by the hydrostatic action of the blood pressure – say 130 mm – which drives the water across. But if the pressure falls below 30 mm, because of disease or shock, urine ceases to be formed.

2 The ordinary laws of *diffusion* and *osmosis*, depending on the relative strengths of the solutions on either side.

3 Differences of *electrical potential*.

4 The varying *permeabilities* of the membranes. They are only rarely totally impermeable, and none is completely semi-permeable, allowing passage of water only. Some allow certain colloids to cross, more often only water and crystalloids. In some cases substances pass only in one direction, due to changes they produce in the membrane. Finally, in the dialysis of a compound of one diffusible and one non-diffusible ion, a difference of *chemical reaction* develops on the two sides; this accounts for the secretion of the highly acid and alkaline digestive juices of the stomach and pancreas respectively.

Energy transformations within the body

The essential chemical elements of protoplasm are carbon, hydrogen, nitrogen, sulphur and phosphorus. Simple compounds of these, such as water and carbon dioxide, require an elaborate conversion into organic matter which is made by green plants under the influence of sunlight. Neither humans nor other animals can achieve this; they depend on the consumption of vegetable matter, either directly or indirectly after it has been utilized by other animals which serve as food in their turn. Ultimately all flesh is grass (or plankton). The energy transformations of living matter begin with light absorbed by plants and end with the heat waste of plants and animals. This passage is subject to the laws of thermodynamics and can be used to do *work*. During digestion and absorption, the constituents of the food are broken down into simpler organic substances and reassembled in the tissues. The final products differ slightly in composition in humans and other animals, and to some extent between individuals. Humans depend on complex organic compounds – fats, carbohydrates and proteins –for body-building material and energy sources . The only chemical element taken up in the free state is the oxygen of the inspired air. Oxygen frees the potential energy of the food by burning it in the body cells to form carbon dioxide and water, liberating energy. Thus this last state is the complete reversal of what the green plant originally achieved. Since the breakdown products of animal tissues are used again by plants, there is a continuous cycle on which both plants and animals depend. And the energy for our body heat and movement is ultimately derived from the sun.

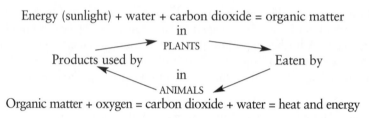

Energy (sunlight) + water + carbon dioxide = organic matter
in
PLANTS
Products used by   Eaten by
in
ANIMALS
Organic matter + oxygen = carbon dioxide + water = heat and energy

The energy concentration of the green leaf is very low, hence the large amounts needed by herbivores. Nevertheless, this energy can be concentrated by herbivores into fat and meat for use in human diet.

This cycle concerns only carbon, hydrogen and oxygen, but similar cycles apply to the other essential elements such as nitrogen. This is mainly synthesized into organic matter by the soil bacteria on which plants depend. The plants pass it on to humans and other animals, which return it in the urea and ammonia of their excreta to the soil. A number of metallic ions essential to the body – sodium, potassium, calcium, magnesium and iron – are taken up in the form of simple mineral salts. Essential basic radicals are phosphates and chlorides.

Essential minerals

Sodium: most natural foods contain little sodium, hence the importance of salt in the diet. Though a healthy person does not retain salt excess, the body content is maintained well above that of the environment by the kidney. Sweat contains salt and severe salt loss may occur in hot or humid conditions and cause cramps.

Potassium is abundant in plant and animal cells and dietary deficiency is rare. Severe potassium loss may occur in vomiting and diarrhoea and causes weakness and thirst.

Calcium is essential for bone formation during growth and the maintenance of bone mass in the adult since there is a constant turnover of the kilogram of calcium contained in the skeleton. Most of the dietary calcium is lost in the faeces as insoluble salts; the problem is one of absorption. Milk and cheese are rich sources. The calcium of bread is more easily absorbed in white bread, as in wholemeal the phytic acid in the husk forms insoluble calcium phytate which is excreted, so it is reasonable to enrich high-extraction bread with calcium. Calcium is necessary for normal electrochemical excitability of tissues; deficiency can cause twitching and spasm of the muscles, known as *tetany*.

Magnesium is the most important intracellular cation after potassium and is essential in certain enzyme reactions.

Iron is an essential mineral as part of the molecule of the red pigment of the blood, haemoglobin, responsible for transferring oxygen from the lungs to the tissues. The body contains 4–5 g of iron, half in the haemoglobin. It is in constant demand for synthesis of the pigment in new red cells. In females, a good deal is lost at each menstruation, and it is needed in pregnancy for the foetus and later for the milk. Hence the iron requirements of a woman of childbearing age are twice those of a man and iron-deficiency anaemia is more common in women.

Most of the dietary iron is lost in the faeces and the amount absorbed is only just adequate for women. Rich sources are offal, meat, fruit and vegetables, and iron can be absorbed from cooking-pots. Oddly, the body is quite unable to excrete iron and excessive intake may lead to iron deposits in the tissues.

There is a group of *trace elements*, which, though present in the tissues in minute quantities, are essential dietary constituents. They include iodine, fluorine, copper, cobalt, zinc and manganese.

Iodine is an essential constituent of the secretion of the thyroid gland, which enlarges as a swelling in the neck, or *goitre*, if the supply is inadequate. Goitres occur in isolated continental areas far from the sea and are prevented by adding potassium iodide to table salt.

Fluorine is necessary for hardening the enamel of the teeth; excess causes brown mottling and increased density of the bones. Where the fluoride content of drinking water is low, dental cavities are more common and can be countered by fluoridation of the water supply.

Copper may be important in the formation of haemoglobin.

Cobalt forms part of the molecule of the B12 vitamin.

Zinc is needed in some enzyme systems and for wound healing.

Manganese is also important for certain enzymes.

Nutrients

These can be grouped into three classes: fats, carbohydrates and proteins.

FATS

Fats are also known as *lipids*. Fat is found in the adipose tissue beneath the skin and elsewhere, as around the kidneys, and as globules in the cells everywhere, particulary in the liver. They are essentially combinations of *glycerol* with three *fatty acids* – *oleic*, *palmitic* and *stearic*. Their relative proportions determine the solidity and melting-point of each fat, e.g. the fluidity of human fat compared with solid mutton fat at body temperatures. Oleic is the most fluid, palmitic and stearic are relatively hard. Fats are insoluble in water, dissolve easily in ether and alcohol, can be emulsified in water as fine dispersed droplets, and break down during digestion into their constituent fatty acids prior to rearrangement in the tissues.

Fats provide the largest source of energy next to carbohydrates and in concentrated form, so they are valuable for manual work and in cold climates. They are vehicles for the *fat-soluble vitamins* A, D, E and K. A distinction is made between *essential fatty acids* known as *polyunsaturated*, which have to be ingested as they cannot be synthesized, and the *saturated* animal triglyceride fats, solid at body temperatures. The unsaturated fatty acids are abundant in vegetable and fish oils. Fats are oxidized to carbon dioxide and water; but if the diet is deficient in carbohydrates, the breakdown may be incomplete and intermediate metabolic products appear in the urine – the condition known as *acidosis* or *ketosis* because of the presence of ketones, as may occur in diabetes.

The grease of the skin, oil of the hair and wax of the ear are complex fats know as *sterols*; and even more complicated combinations with phosphorus are found in nervous tissue and the bounding membranes of cells.

CARBOHYDRATES

These may yield half the total calories in western communities, but some people are healthy with virtually none, so they are not essential. This is because the body can form glucose from fats and proteins. They are the immediate product in plants of the synthesis of water and carbon dioxide and their essential saccharide unit, CH_2O, is the building block used to form compounds of increasing complexity.

The *monosaccharides* may be *hexoses* – $C_6H_{12}O_6$ – such as glucose, also formed in the body by digestion of cane sugar and starch, or *pentoses* – $C_5H_{10}O_5$ – the *deoxyribose* and *ribose* of DNA and RNA respectively. *Galactose* and *fructose* are isomers of glucose, with the same atoms but arranged differently in space.

The *disaccharides* of general formula $C_{12}H_{22}O_{11}$ include *cane sugar*, the *maltose* of fermenting grain, and the *lactose* of milk.

The *polysaccharides* are complex polymers of high molecular weight. They include *starch*, found as grains in plant cells and digested into glucose. *Glycogen*, or animal starch, is very important and is stored as granules in all the tissues, particularly the liver and muscles. The liver glycogen is the great energy reserve of the body, mobilized as required by conversion into glucose. The rough *cellulose* of plants is not digestible to any useful extent by humans but its bulk encourages movement of faeces along the colon.

PROTEINS

The proteins are the most important constituents of protoplasm. They occur in quantity in lean meat, cheese and the vegetable pulses, and are very complex unions of carbon, hydrogen, oxygen and nitrogen, usually also of sulphur. They are colloids, soluble in water and coagulating on heating. The essential unit of the protein molecule is the *amino acid*, containing an amine group – NH_2 – and an organic acid group – COOH. There are some 20 common types. The linkage of one amino acid with another is called a peptide bond and proteins are *polypeptides* containing more than 50 amino acids, up to over 10,000. Though there are thousands of proteins in the body, with widely differing functions, all are varying combinations of the original 20 amino acids. The only chemical differences in the proteins are the particular amino acids of the molecule, their properties, and the arrangement of their peptide linkages. Physically, their molecules may be varyingly twisted or folded or globular. A protein loses its characteristic shape and function when it is *denatured* by heat or acidity; up to a point this is reversible, later it is irreversible, as in the coagulation of egg white.

Different tissues have their characteristic proteins: the red *haemoglobin* of blood, the slimy *mucoprotein* of mucus, the *casein* of milk, egg *albumin*, the phosphorus-containing complexes of *nucleoproteins*. Proteins can function as both weak acids and weak alkalis and thus maintain a buffering neutrality in the tissue fluids.

The protein of each animal species is unique. The reaction a foreign protein excites in the tissues causes the phenomena of the *immune response* and *allergy*.

The proteins are not completely broken down in the tissues and much of their nitrogen is lost in the urine as urea, uric acid and so on. Only 70% of the energy theoretically available in protein is actually utilized by the tissues, whereas with fats and carbohydrates, breakdown is total, and the energy yield is exactly what would have been obtained by combustion in the laboratory. *Nitrogen balance* is between daily intake and daily loss in the urine. It is negative in starvation and after major injuries or operations. It is positive, i.e. there is nitrogen retention, in growing children and during convalescence for patients on an adequate diet.

Within limits fats and carbohydrates are interchangeable as energy sources, but the diet *must* contain protein to supply nitrogen and certain 'essential' amino acids that the body cannot synthesize. These are found only in the 'first-class proteins', such as meat. In western countries, about 10% of energy intake is provided by protein, 40% by fat and 50% by carbohydrate (see also Chapter 13).

Enzymes

The complex changes of chemical breakdown and resynthesis are performed much more rapidly in the body than in the laboratory. Starch is converted into maltose by the amylase of the saliva in a minute, but this takes the chemist several hours' boiling to achieve.

This facilitation is due to *enzymes*. We are familiar in inorganic chemistry with *catalysts* which are only needed in minute amounts to accelerate reactions in which they are not themselves consumed. Enzymes are the catalysts of the organic processes of the body. They are complex colloidal materials, globular proteins preferring a definite degree of acidity or alkalinity for optimum performance, e.g. the pepsin of the acid gastric juice, the trypsin of alkaline pancreatic secretion. They facilitate breakdown, synthesis and oxidation processes, but especially the process of *hydrolysis*, when water is added to a molecule, often as a preliminary to its disintegration.

Enzymes are *specific*, each acting only on one substance, or *substrate*, e.g. the amylase of saliva acts only on starch; and the digestive juices contain different enzymes for the various constituents of the food – lipases for fats, proteases for proteins, carbohydrases for sugars and starches; and they do not act on other substances. Most enzyme reactions are *reversible* and fall into two main groups: digestion within the bowel, and the more intimate reactions, such as oxidation in the individual cells.

Most metabolic reactions occur only in the presence of an appropriate enzyme, and the rate of reaction is related not only to the concentration of substrate and enzyme, local pH and temperature, but also to the presence of certain *coenzymes* and *activators*, which can be metallic ions such as zinc or magnesium or organic molecules. As proteins, enzymes can be *denatured* by heat. Many have been isolated in pure or crystalline state. They are classified by their effects, as follows:

1 *oxidoreductases* in oxidation/reduction processes

2 *transferases* in transfer of chemical groupings from one molecule to another

3 *hydroxylases* in hydrolysis of compounds into simpler molecules

4 *lyases* in splitting compounds

5 *ligases* for joining segments of molecules

6 *isomerases* in rearrangement of molecules to form isomers.

Metabolism

Physiologically, human life can be regarded as a continuous production of energy by oxidation (burning) of food, and expenditure of this energy in:

1 maintaining a body temperature above that of the surroundings

2 movement.

The balance sheet for this energy intake and output is exact; food burnt in the body liberates the same energy as when oxidized in the laboratory. However, the human arrangements are complicated because the food is not burnt *as such*, but only after it has been digested and assimilated into the living tissues. The fuel has to become part of the furnace. Only a few substances like alcohol can be burnt directly without first becoming part of the living protoplasm.

Thus there is a recurrent cycle of activities. Part of this is the process of building-up or repair, of assimilation of digested food into the tissues – *anabolism*; part is the breaking down of these tissues with the liberation of energy and excretion of wastes – *catabolism*. Both are going on all the time, though their relative proportions vary: anabolism preponderates during growth; catabolism during starvation and in senescence.

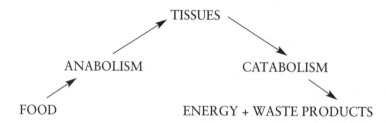

The wastes are excreted by the kidneys, bowel, lungs and skin. They consist mainly of water, carbon dioxide and nitrogenous breakdown products of protein such as urea, which are mainly expelled in the urine.

During fasting, at complete rest, there is still an essential minimum energy output to maintain warmth and the movements of respiration and the heart. This minimum can be measured and is known as the *basal metabolic rate*. Although of the same order of magnitude for everyone, it varies from person to person by virtue of their different sizes. This is because most of the energy is expended in making up for heat loss from the body surface, and is therefore proportional to surface area. Basal metabolism is higher in men than in women and higher still in young children. Each individual has his or her normal metabolic rate, predictable from height and weight, and this is increased in fever and excitement and also by hyperactivity of the thyroid gland. In thyroid deficiency, metabolism is considerably below normal.

From the basal rate we can estimate the calorie value of the food required to maintain essential activities of the body. This is around 2000 calories per day for an adult male. Below this level the body begins to fall back on its reserves. The fat and glycogen stores are first consumed, and then the less essential tissues, such as the muscles and glands, digest themselves to feed the brain and heart. This will ultimately lead to death due to heart failure.

Test yourself

1 What is the normal pH of blood?
   a   7
   b   7.2
   c   7.4
   d   7.6

2 Which receptors detect when excess water has been consumed?
   a   Chemoreceptors
   b   Osmoreceptors
   c   Exteroreceptors
   d   Proprioreceptors

3 What is the total body water in a normal healthy adult male?
   a   20 litres
   b   30 litres
   c   40 litres
   d   50 litres

4 Which statement describes the process of osmosis?
   a   Water is drawn from the weaker to the stronger solution through a semi-permeable membrane
   b   With two solutions of different strengths, their concentrations are equalized by the transfer of water and solute in reverse directions
   c   It is the energy gradient set up by the tendency of water always to dialyse so as to dilute the stronger solution
   d   The flow of gas from a high to a low concentration

5 Which of the following is a crystalloid?
   a   Water
   b   Gelatin
   c   Egg white
   d   Salt

6 What is the function of iron in the diet?
   a   Hardening of the enamel of teeth
   b   Needed for wound healing
   c   Part of haemoglobin
   d   Essential in certain enzyme reactions

7 When is there a negative nitrogen balance in the body?
   a   In growing children
   b   During convalescence
   c   After a major injury
   d   During pregnancy

8 What does the term 'adsorption' mean?
   a   When gases flow from a high to a low concentration
   b   Picking up and retaining a substance without entering into a chemical combination with it
   c   Physical factors that control the transference of substances across cell membranes
   d   A solution with high osmotic pressure

9   What is the function of an enzyme?
    a   To speed up the rate of a reaction
    b   To maintain nitrogen balance in the body
    c   To provide energy for metabolism
    d   To provide the body with mineral salts

10 Which person would have the relative highest basal metabolic rate?
    a   A 30-year-old adult male
    b   A 30-year-old adult female
    c   A 6-year-old child
    d   A 70-year-old male adult

13

Foodstuffs and vitamins

In this chapter you will learn about:

▶ *food and energy*

▶ *diet*

▶ *respiratory quotient*

▶ *food groups*

▶ *vitamins.*

Food and energy

Energy from *light* is absorbed by plants for photosynthesis and initiates every biological process. The conversion of light energy permits useful *work*. The bound chemical energy in working muscles is transformed partly into work, partly into heat. The basic physical unit of heat is the *calorie*, the amount required to raise the temperature of 1 gram of water from 14.5°C to 15.5°C, equivalent to 4.2×10^7 ergs. (Under the SI system, 1 kcal (1000 cal) = 4.186 megajoules.)

The fuels of the body are protein, fat and carbohydrate. Most of the energy in their carbon and hydrogen linkages can be made available by oxidative metabolic processes. The heats of combustion of these foods measured in the laboratory are slightly higher than the energy actually available to the body, since digestion and absorption are incomplete and oxidation is only partial. Nevertheless, 99% of the carbohydrate, 95% of the fat and 92% of the proteins taken are absorbed; only 5% of the calorie intake is lost in the faeces. Since much of the nitrogen intake is excreted in the urine as urea, uric acid and other substances, about a quarter of the calorie value of ingested protein is lost. The amounts of heat available from 1 gram of a basic foodstuff are:

▶ 1 g of protein   4 kcal

▶ 1 g of fat   9 kcal

▶ 1 g of carbohydrate   4 kcal

It is possible to calculate the energy value of a diet by analysing its components and applying these factors (see Table 13.1).

These foods not only supply energy but also replace wear and tear in the tissues. Only proteins can be used for cell building.

Fat is used for energy production as it is more economical than the others, since it provides twice as much heat and is used in a much purer form.

Proteins are not an economical way of obtaining energy; fats and carbohydrates are normally used to meet energy requirements and proteins to repair tissue breakdown. A man would have to eat 2 kg of lean meat daily to obtain his basic minimum of 2000 calories, an amount impossible to digest.

Nor is protein stored to any extent, except during growth. Protein, though an essential of the diet, has no advantages as an energy source over fats and carbohydrates, which are capable of considerable interchangeability.

Diet

The average daily requirement of a working man is 3000 calories. An adult woman needs 2500 calories because of her smaller size; and sedentary workers need less. To maintain health a diet should contain protein, fat and carbohydrate in the proportions indicated below, enough fresh food to supply the necessary vitamins, enough mineral content, and it must be *palatable*.

Protein should supply one-sixth of calorie needs, i.e. at least 100 g a day, and at least half in the form of first-class animal protein; but it is relatively expensive. An excess serves no

Table 13.1 Energy value and nutrient content of common foods (per gram)

	Water %	Protein %	Fat %	Carbohydrate %	Energy %
Wheat flour	15.0	13.6	2.5	69.1	3.4
White bread	38.3	7.8	1.4	52.7	2.4
Rice	11.7	6.2	1.0	86.8	3.6
Milk	87.0	3.4	3.7	4.8	0.7
Butter	13.9	0.4	85.1	trace	7.9
Cheese	37.0	25.4	34.5	trace	4.3
Steak	57.0	20.4	20.4	nil	2.7
Haddock	65.1	20.4	8.3	3.6	1.8
Potatoes, raw	80.0	2.5	trace	15.9	0.7
Peas	72.7	5.9	trace	16.5	0.9
Cabbage, boiled	96.0	1.3	trace	1.1	0.1
Orange	64.8	0.6	trace	6.4	0.3
Apple	84.1	0.3	trace	12.2	0.5
Sugar	trace	trace	trace	100.0	3.9
Beer	96.7	0.2	trace	2.2	0.3
Spirits	63.5	trace	nil	trace	2.2

particular purpose, even for heavy workers. What protein does, more than other foods, is to give a fillip to metabolism generally, which increases heat production and maintains body warmth in cold weather.

Carbohydrates form the bulk of most diets; they are cheap and easily obtainable as bread, rice and other foods. The optimum daily need is 500 g. There is no absolute minimum carbohydrate requirement; however, an intake below 500 g means that the carbohydrate/fat ratio is disturbed. When this happens, fats are not completely burnt and acid products accumulate in the tissues (*ketosis, acidosis*).

Fat is valuable because of its high energy value. A normal diet contains only one-seventh of its weight of fat (some 100 g), but this supplies a quarter of the total calories because of the small amount of water it contains compared with other foods. Although interchangeable with carbohydrate as an energy source, fat has certain advantages. Its slow absorption prevents the hunger felt with a largely carbohydrate diet; and for really heavy work in cold or Arctic conditions, a large amount of fat is the only practical way of supplying calorie needs.

Respiratory quotient

We can deduce which substance is being oxidized in the body to provide energy by comparing the oxygen (O_2) used up (contained in the inspired air) with the carbon dioxide given off (in the exhaled air).

When carbohydrates are burned, the equation is:

$$C_6H_{12}O_6 + 6O_2 = 6CO_2 + 6H_2O$$

And the *respiratory quotient* is $\dfrac{CO_2 \text{ expired}}{O_2 \text{ inspired}}$ which works out at exactly 1.

Since different materials are burnt at the same time, the actual quotient depends on the relative proportions of the foodstuffs, whose respective quotients are:

Carbohydrates	1.0
Protein	0.8
Fat	0.7

The actual quotient is always less than 1, since a purely carbohydrate diet never exists, and it represents the gross results of oxidation of the particular mix of foodstuffs.

For energy any of the foodstuffs may be utilized, and carbohydrates and proteins are called *isodynamic* because an equivalent weight of each produces the same amount of heat. Carbohydrates and fats in the proportion of rather more than 2:1 also provide the same number of calories and can replace each other in these proportions without altering the available energy. Although protein is theoretically interchangeable in the same way, an upper limit is set by the capacity to digest it. Since protein is not stored, any excess is broken down and excreted as nitrogenous compounds in the urine and faeces, i.e. there is a normal state of nitrogen equilibrium. Any excess of either fats or carbohydrates is stored in the body as fat.

Protein is broken down by digestion into its constituent amino acids, which enter the bloodstream and into equilibrium with the protein of the body cells, the latter taking up such as they require for repair. Residual amino acids are disintegrated in the liver; their nitrogenous part is lost as urea and the rest burnt to carbon dioxide and water. Only the *first-class proteins* of meat and wheat and some other sources contain those essential amino acids which cannot be synthesized in the body. The *second-class proteins*, like gelatin, lack some essential amino acids or are composed of inessential ones. Deficiency of essential acids results in loss of appetite and slowed growth; for their provision animal protein is more economical in bulk than vegetable, though the latter may suffice if the diet is varied.

Food groups

Foods may be grouped as follows: cereals; starchy roots; pulses and legumes; vegetables and fruits; sugars and derivatives; meat, fish and eggs; fats and oils; and beverages. No single one is essential, but a good diet contains foods of all classes.

▶ Cereals

These form the chief food of most human beings; even in rich countries they provide a quarter of the total calories and are the largest item of diet. All cereal seeds have much the same structure (Figure 13.1). The grain contains 80% starch, a little fat and as much as 10% of protein; but the protein of any one cereal is poorly suited to human requirements, so a mixture of sources is required. Cereals contain much calcium and iron. Whole grains are a valuable source of B vitamins. They contain no vitamin A, C or D; but oily grains like maize contain vitamin A precursors and much vitamin E.

Wheat and *rice* are usually eaten after milling to white flour and polished rice. The material removed contains much of the B vitamins and thiamine, so rice-eaters may suffer from *beri-beri* (see below). The losses can be made good by enriching white flour with B vitamins; however, it is difficult to enrich polished rice.

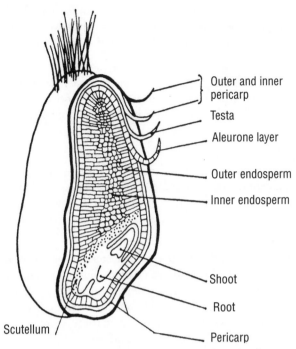

Figure 13.1 The structure of the wheat grain (from *McCance, Lancet i* (1946), 77)

The poor man's cereal is *millet*, which is not highly extracted. *Maize* is an excellent food but if it forms the bulk of the diet it may be associated with pellagra (see below). The yellow form is better than the white.

The nutritive value of cereals is considerable. Bread and rice provide calories, proteins, B vitamins, calcium and iron. Only when a diet consists of an excess of cereals, or of any one cereal, does it become unbalanced.

▶ Starchy roots

The *potato* is an excellent, energy-rich food. On no other single food can a human survive indefinitely, for potatoes contain significant amounts of vitamin C and carotene and adequate amounts of the B vitamins. Tropical roots are the *cassava* and *yam*, a rich source of energy, though their protein content is not high. If children are fed largely on these roots, they are liable to protein deficiency diseases.

▶ Pulses

Pulses, e.g. *lentils* and *dhals*, have a protein content double that of cereals and are valuable in the tropics. They are also rich in B vitamins. The *soya bean* contains up to 20% fat.

▶ Vegetables and fruits

Vegetables include leaves, flowers, seeds, stems and roots. They contain much water, few calories and little protein. These types of food are filling and prevent constipation due to their fibre content. They contain some vitamin B and C, but the latter is often lost in cooking. The chief vitamin is *carotene*, the precursor of vitamin A. Vegetables increase the nutritive value of cereal diets in the tropics but contribute little to the average western diet, though they have been linked to providing some degree of protection against cancer. They also contain folic acid (see below).

The main nutritional value of *fruits*, especially the citrus fruits, lies in their vitamin C content. The banana provides carbohydrate in abundance but lacks protein. Nuts are rich in fats and carbohydrates; they also contain protein, and manganese and other trace elements.

▶ Sugar

There has been a great increase in sugar consumption in the last 150 years and this has been linked to an increase in dental cavities and cardiovascular disease. Without other nutrients, sugar provides 'empty' calories and promotes obesity and diabetes.

▶ Meat, fish and eggs

Meat, with its fat, contains one-fifth fat, three-fifths water and one-fifth protein of high biologic value. It is also a source of iron and B vitamins, but not of vitamins A or C. *Offal*, such as liver and kidneys, has the same general value as meat but more vitamin A and iron. *Fish* is a rich source of protein, unsaturated oil (which has a cardioprotective action), and vitamins A and D. *Eggs* are a most useful item of a diet as they consist of 10% first-class protein, 10% of fat, much vitamin A, and calcium and iron.

▶ Milk

Human milk is a complete infant food. Cow's milk contains more protein and less carbohydrate, but can be made suitable by dilution and adding sugar.

Table 13.2 Milk values (per 100 grams)

	Human milk	Cow's milk
Calories	70	66
Carbohydrate	7 g	5 g
Protein	2 g	3.5 g
Fat	4 g	3.5 g
Calcium	25 mg	120 mg
Phosphorus	16 mg	95 mg
Iron	0.1 mg	0.1 mg
Vitamin A	170 units	150 units
Vitamin C	3.5 mg	2.0 mg
Vitamin D	1.0 units	1.5 units
Thiamine	17 μg	40 μg
Riboflavine	30 μg	150 μg
Nicotinic acid	170 μg	80 μg

Milk is rich in calcium but poor in iron, and infants may develop anaemia after six months if mixed foods are not added. Boiling milk destroys vitamin C, so scurvy may develop. Mothers usually produce adequate milk of good quality, even at their own expense. Milk is probably essential for growing children, but not for adults. *Cheese* is made by clotting milk with rennet; it retains most of the original protein and fat (and cholesterol) and is highly nutritious.

▶ Fats and oils

These may be animal or vegetable. The former are mainly dairy fats, beef suet and pork lard; the latter include oils of the olive, groundnut, cotton seed and red palm and corn oil. Margarine is made by mixing vegetable oils. Animal fats (except those of fish) are solid and rich in saturated fatty acids, which have acquired a bad reputation in cardiovascular disease. Vegetable oils are liquid and contain a lot of unsaturated fatty acids, as does fish oil. Fats are a concentrated source of calories, keep the diet from being too bulky, and are vehicles for the fat-soluble vitamins. Butter contains vitamins A and D, which have to be added to margarine, but there is little to choose between the two in nutritive value. Vegetable oils contain no vitamins.

▶ Beverages

These are drunk for their flavour or alcohol content. *Ethyl alcohol* is produced by fermenting grain or fruit. It is a nutrient, readily absorbed, and an immediate source of energy. It is a sedative and promotes the feeling of well-being in moderation. Some people can become addicted to alcoholic drinks, and alcohol has toxic effects on the liver and nervous system if taken in excess over long periods.

Beer is made by converting grain starch to maltose by maltase and then fermenting it by yeast. Beer contains often around 6% alcohol and the vitamin riboflavine. *Wine* is made by fermenting grapes or other fruits; it contains around 8–10% of alcohol and may be fortified with added spirits up to often around 15%. It may contain large amounts of iron, which can cause chronic liver disease. *Spirits* are made by distilling fermented liquor; they may contain often around some 30% of alcohol but are devoid of other nutrients. Imperfect distillation may yield spirits containing *methyl alcohol*, which is very poisonous and can cause blindness or death.

Coffee and *tea* contain *caffeine, theophylline* and *theobromine*. Caffeine stimulates the nervous system and increases the output of urine. Both contain *tannins*, which are protein precipitants and may cause gastritis. Although tea and coffee are of no nutrient value, they are usually taken with milk and sugar and this may be a hidden source of calories.

Soft drinks may contain vitamin C; their calorie value depends on the added sugar. *Fruit juices* are rich in potassium and useful to convalescents. *Mineral waters* have a varied but insignificant mineral content.

Vitamins

Natural foods contain minute amounts of organic substances known as *vitamins* that the body is unable to synthesize (except, to some extent, vitamin D). Many are components of essential enzyme systems. They are essential for the proper utilization of food, normal growth and to maintain health in adults. Yet they contribute nothing towards energy requirements or tissue repair. There are five major diseases due to lack of specific vitamins: scurvy, beri-beri, pellagra, keratomalacia (ocular disease) and rickets.

Vitamins are either *fat-soluble* or *water-soluble*. The water-soluble vitamins cannot be stored as they are rapidly excreted by the kidneys, so that deficiency disease soon arises if the intake is inadequate. The fat-soluble vitamins (A, D, E and K) are stored in adipose tissue, particularly the liver, and a well-fed individual will remain healthy for many weeks with their own stores of these vitamins even if the external supply is cut off.

FAT-SOLUBLE VITAMINS

Vitamin A (retinol) is present only in certain animal foods – dairy produce, liver and particularly fish-liver oils, although it was originally formed in green plants. The small intestine can also manufacture vitamin A from the carotene pigment consumed from some plant sources. In the UK half the dietary vitamin A is obtained from animal sources and half from precursors in plants. Deficiency disease of vitamin A is rare in the UK, but it is common in the tropics. Deficiency of this vitamin causes stunting, susceptibility to disease, defective repair of epidermal tissues, and progressive collapse and disorganization of the eye (*keratomalacia*).

Vitamin D (calciferol) is found in dairy produce and in eggs and liver. It is also formed in the skin from the action of ultraviolet light on a steroid precursor, so the body is not entirely dependent on dietetic sources. However, heavily pigmented children from the tropics can develop rickets, which results from a deficiency of vitamin D, when they are moved to a cloudy temperate zone.

Vitamin D promotes absorption of calcium from the bowel and deposition of bone by formation of calcium salts. It has a major influence on the growth and hardening of bones and teeth. *Rickets* is a disease of children in which soft, poorly calcified, osteoid tissue is laid down instead of normal bone, so that deformities develop – spinal curvature, bow-legs or knock-knee, or pelvic deformity hindering later childbirth. It is readily cured by moderate intake of the vitamin. There is an adult form of rickets called *osteomalacia* – usually occurring in older women, especially dark-skinned, whose intake is low.

Vitamin E is an antioxidant tocopherol essential to normal fertility, and research suggests that it is probably also cardioprotective. It lengthens blood-clotting time. Deficiency is very rare in humans, but rats fed on artificial diets may abort, the males become sterile, and embryos fail to develop. The vitamin is found in quantity in vegetable and fish oils and wheat-germ.

Vitamin K is a complex of naphthoquinone derivatives mostly produced by bacteria in the large intestine but also present in green vegetables. Essential to the complex mechanism of blood clotting (see Chapter 16), it can be synthesized. Dietary deficiency does not exist.

WATER-SOLUBLE VITAMINS

Vitamin B is a complex of three factors.

1   *Vitamin B1* (*thiamine*) is abundant in the husks of cereal grains, pulses, yeast, milk and meat; it is lost when wheat and rice are subjected to machine milling. Thiamine catalyses the oxidation of carbohydrates in cells, particularly nerve cells, and a deficiency causes beri-beri. In *dry beri-beri* there is peripheral neuritis, muscle wasting and paralysis; in *wet beri-beri* there is heart failure and waterlogging of the tissues. Wherever rice is a staple and finely milled, there is a risk of the disease and, even in western countries, synthetic thiamine is often added to white flour. The required daily intake is about 1.5 mg.

2   *Nicotinic acid* is another member of the B complex, with multiple animal and vegetable sources and easily synthesized from proteins. Deficiency causes *pellagra* – dermatitis, dementia and diarrhoea – associated with the exclusive consumption of maize, which is a poor source. But the body can synthesize nicotinic acid from the tryptophane in rice, so that rice-eaters are protected even though rice itself contains little of the natural factor. Pellagra may also result from interference with intestinal absorption, as in alcoholism. Nicotinic acid is present in whole cereals and rice, meat, offal and fish, and it is abundant in yeast and yeast extracts. It is not present in in dairy products, fruit or vegetables. In the UK and USA

it is legally required to be added to white flour. The mean necessary daily intake is of the order of 10–12 mg.

3 *Riboflavine* is a yellow pigment, an essential coenzyme for tissue respiratory processes. It occurs in most foods. Lack of it may case ulceration at the angles of the mouth and blurred vision. The daily requirement is 1–2 mg.

The anti-anaemic B vitamins: red blood cells only live for 100 days or so and must be continuously replaced. Certain vitamins are essential to this process, and if they are lacking there will be anaemic.

Vitamin B12 (cyanocobalamin) occurs in animal foods but not plants. It is a complex porphyrin derivative containing cobalt, also a by-product in the manufacture of certain fungal antibiotics. The daily requirement is less than 1 µg. For proper absorption from the bowel, an *intrinsic factor* secreted by certain stomach glands is essential; when this is absent the individual develops *pernicious anaemia*. This may be cured by eating large amounts of liver, or more usually by regular injections of the vitamin. *Folic acid* is the other anti-anaemic B vitamin, whose main source is the leaves of green vegetables. It is essential for normal development of the spinal cord, and deficiency may be a cause of spina bifida; also for formation of red blood cells. The average daily requirement is some 50 µg. Anaemia due to deficiency is common in tropical countries, particularly in pregnancy.

Vitamin C (ascorbic acid) is a crystalline sugar, readily synthesized, soluble and easily oxidized. The main source is from fruit, vegetables and milk. Meat and eggs contain only a trace; potatoes are a useful source. It is easily destroyed by cooking, and when diets contain little fruit or vegetables, or vegetables are boiled, *scurvy* may result. Scurvy is characterized by haemorrhages such as bleeding gums, and indolent healing of wounds and fractures. This can be easily prevented by consuming citrus fruits. Milk contains little ascorbic acid and what exists is destroyed by boiling, so infants are at risk. The minimum daily intake is about 25 mg, provided by an ordinary helping of cabbage or other green vegetables, a good helping of potatoes twice daily, or by 40 g of fresh oranges or orange juice. It cannot be stored to any useful extent in the body; when excess is consumed, it is passed in the urine, so regular intake is important.

To sum up, most vitamins are of known composition and can be synthesized or obtained in pure form. The only evidence of their effects is that when the body is deprived of them, deficiency diseases appear. Although essential to health, they are not foodstuffs and cannot compensate for an inadequate diet. Further, in adults living on a good mixed diet, there is no need for vitamin supplements. This is not always so with growing children, where supplements of fish oil and orange juice may be desirable. Chronic alcoholics, because their drug interferes with absorption and storage of vitamins, may need large doses of multi-vitamin preparations.

Test yourself

1 What percentage of calorie intake should come from proteins?
- **a** 9%
- **b** 11%
- **c** 14%
- **d** 17%

2 Which foodstuff has the highest amount of heat available from 1 g?
- **a** Fat
- **b** Carbohydrate
- **c** Protein
- **d** Vitamins

3 Which food contains the highest amount of carbohydrate?
- **a** Red meat
- **b** Rice
- **c** Milk
- **d** Carrots

4 What is the equation for working out the respiratory quotient?
- **a** CO_2 expired/O_2 expired
- **b** O_2 inspired/CO_2 expired
- **c** O_2 expired/ CO_2 inspired
- **d** CO_2 inspired/O_2 expired

5 Identify a food that contains essential amino acids.
- **a** Potatoes
- **b** Gelatin
- **c** Meat
- **d** Cheese

6 Which food group provides a quarter of total calories for most humans?
- **a** Cereals
- **b** Starchy roots
- **c** Fruits
- **d** Vegetables

7 Which type of food is from an animal source?
- **a** Red palm oil
- **b** Pork lard
- **c** Cotton seed oil
- **d** Ground nut oil

8 Which of the following vitamins are water-soluble?
- **a** Vitamin A
- **b** Vitamin C
- **c** Vitamin D
- **d** Vitamin K

9 Which disease is linked to vitamin D deficiency?

    a   Scurvy

    b   Beri-beri

    c   Anaemia

    d   Rickets

10 Which vitamin is required for normal development of the spinal cord of a foetus?

    a   Carotene

    b   Nicotinic acid

    c   Riboflavine

    d   Folic acid

14

Digestion

In this chapter you will learn about the process of digestion and the digestive tract, including:

▶ *the mouth*

▶ *swallowing*

▶ *the stomach*

▶ *the small intestine*

▶ *the pancreas*

▶ *the liver and bile*

▶ *the large intestine*

▶ *movements in the digestive tract.*

The alimentary tract is a long muscular tube extending from mouth to anus, lined with an epithelium which secretes digestive juices and absorbs the products of digestion. There are also major glands lying outside the bowel altogether – the salivary glands, liver and pancreas – whose secretions are discharged into the intestine. These secretions amount to many litres in 24 hours and most of this fluid is reabsorbed, so vomiting or diarrhoea may cause severe dehydration and electrolyte disturbance. Secretion takes place mainly in the upper part of the tract, absorption in the middle, and excretion in the lower segment.

Digestion and absorption are unconscious processes. Bowel movement is involuntary and effected by smooth muscle; only chewing, the initiation of swallowing and defaecation are conscious processes.

The *object of digestion* is to convert complex and insoluble food constituents into simpler substances that can dissolve and diffuse through the lining of the intestine to enter the blood or lymph. Thus, carbohydrates must be reduced to the simplest monosaccharide form or the cells will not be able to assimilate them. Fats must be hydrolysed to fatty acids and glycerol, although some fats enter the lymph unhydrolysed as finely emulsified globules. Proteins are split into their constituent amino acids. In dealing with these the body displays considerable versatility; to convert the protein of cheese into a human form of protein is like taking a construction toy model to pieces and picking out the right bits to build a new one. Digestion is accomplished by the enzymes in the juices formed in the glands along the alimentary tract. Each foodstuff needs a specific enzyme to break it down.

The mouth

Entry of food into the mouth promotes a reflex flow of saliva. The stimulation of the taste nerve-endings is transmitted to the brain and relayed along the nerves controlling salivary secretion. But there may also be secretion on mere anticipation, the sight or smell of food, and this is a conditioned reflex in which an association of ideas has the same effect as the physical stimulus.

The *saliva* is a mixture of the secretions of three pairs of salivary glands – the *parotid* in the cheek, the *submandibular* below the angle of the jaw, and the *sublingual* beneath the tongue – as well as of many minute glands in the mucous membrane of the mouth and palate. Saliva contains *mucin*, which makes the food easy to swallow; *amylase*, an enzyme that digests starch to maltose; and *minerals*, particularly calcium phosphate, which is deposited on the teeth as tartar. Much of salivary digestion actually occurs in the stomach, since food does not stay long in the mouth and the swallowed mass still contains amylase, which continues to act until penetration of the bolus by the acid gastric juice brings it to a halt. There is normally a continuous slight secretion of saliva which is suppressed by fear and by drugs that dry the mouth. No absorption of nutrients occurs in the mouth. The saliva also combats noxious bacteria, partly due to the liberation of nitric oxide from food nitrates by friendly bacteria at home on the tongue.

Chewing (mastication) breaks up the food and mixes it wth saliva, and small amounts are passed at intervals into the pharynx to be swallowed. Swallowing is initially a voluntary activity but involuntary movements soon take over.

Swallowing

The oesophagus is a long muscular tube, beginning behind the lower part of the larynx and travelling down the back of the thoracic cavity to pierce the diaphragm and join the stomach. Chewing produces a round *bolus* of food which is passed down to the stomach in the act of swallowing. In *swallowing* it is essential to prevent food passing upwards into the nose round the back of the soft palate, or down the trachea into the lungs. Therefore the soft palate is apposed to the back wall of the pharynx to seal off the nasal cavities. The epiglottis is an elastic flap of fibrocartilage situated behind the root of the tongue which bends backward to close off the larynx. At the same time, the larynx as a whole moves upward and then descends again (Figure 14.1). The initiation of swallowing is largely voluntary, but once the food touches the pharynx, soft palate and epiglottis, the process continues as a reflex. In the oesophagus, the food is propelled onward by *peristalsis*, a wave of contraction controlled by the nerve plexuses in the oesophageal wall. At the lower end, the cardiac sphincter at the stomach entrance relaxes reflexly and food enters the stomach. This sphincter is normally closed to prevent reflux of acid gastric juice into the lower oesophagus, which may cause serious irritation or even overflow into the airway. The mean *transit time* along the oesophagus is some two seconds for fluids, four to eight seconds for solids.

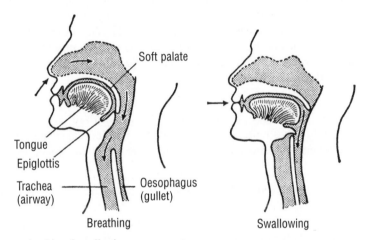

Figure 14.1 The mechanism of swallowing

Note how the palate moves upward to block the back of the nose and the epiglottis down to shut off the windpipe

The stomach

The stomach is a hollow muscular organ with a glandular mucous membrane which secretes the gastric juice. It acts as a reservoir in which a mass, known as the *chyme*, is formed from the ingested food. Digestion continues until the chyme is transferred to the duodenum, but there is little absorption here except for water, alcohol and simple crystalloids.

Small amounts of gastric juice are present in the stomach in resting state. The anticipation of food or its entry into the mouth excites a reflex or conditioned secretion. More is formed when food actually enters the stomach. A pint of highly acid juice may be secreted after a heavy meal and the process continues for up to four hours. The stimulus to secretion of gastric juice is from a hormone, *gastrin*, which is released into the circulation from the pyloric region of the stomach and stimulates the body of the organ. Protein, alcohol and coffee are potent excitants of this phase. The secretion has up to 0.5% of hydrochloric acid, some of which is neutralized by mucus; only some 0.3% is present in the free state. The passage of hydrogen ions into the cavity is necessarily accompanied by an increase of negative ions, mainly bicarbonate, in the cells. So the blood leaving the stomach is more alkaline than that arriving, but this is neutralized by the acid blood leaving the pancreas.

The enzyme present in the stomach is *pepsin*, which converts proteins into polypeptides; in children, *rennin* is also present in the stomach, and curdles milk into a casein clot to be digested by pepsin. Gastric juice acts by virtue of its acidity and through its enzyme action. The acid alone is responsible for minor breakdown of certain foods, but most of the work is done by the pepsin, which also helps liberate the fat from the food by dissolving the fibrous framework round its globules, preparing it for digestion in the duodenum. There is also the *intrinsic factor* required for absorption of vitamin B12, essential for red cell production; its absence causes pernicious anaemia.

The empty stomach is not entirely relaxed; small contractions travel from the cardiac end toward the pylorus. A meal distends the organ and this excites a series of contraction waves which propel the food from the fundus to the pylorus on the right. For the first 30–60 minutes after a meal, the opening into the duodenum at the pyloric sphincter remains closed. Thereafter, it opens at intervals and allows gastric contents to pass in intermittent spurts into the duodenum. By the end of gastric digestion, in 3–4 hours, all the food has passed, the sphincter relaxes, and some duodenal contents and bile regurgitate into the stomach. Fluids leave the stomach long before solids and may enter the duodenum in a few minutes. Fats remain much longer than carbohydrates.

The stomach is not perhaps a very important organ in digestion and absorption, as digestion can take place efficiently without it. Its chief role appears to be to receive and store food at irregular intervals, and to pass more regularly partly digested food in smaller quantities to the small intestine, where most digestion occurs. However, the intrinsic factor is produced nowhere else.

The mobility of the muscular wall and the rate of secretion are intimately affected by emotional states, so that it is not surprising that dyspepsia or ulceration are often of nervous origin. However, it has recently been shown that the ultimate factor in ulceration in the stomach or duodenum is infection with the bacterium *Helicobacter pylori*.

Small intestine

It is here that most of the digestive process occurs, effected mainly by the secretions of the liver and pancreas, which discharge into the duodenum. The principal function of the bowel wall is absorption. The three digestive juices are the *pancreatic juice*, the *bile* formed by the liver, and the *intestinal juice* produced by the mucous lining of the bowel. The bile and pancreatic juice are discharged into the duodenum via ducts from the pancreas and liver which open in common. The intestinal juice is formed along the entire length of the small intestine but plays no great part in the process. It does contribute water to aid solution and absorption of digested food. Secretion and digestion predominate in the upper coils of the small bowel and absorption of digested products into the portal circulation in the lower.

Pancreas

This organ secretes the pancreatic juice into the duodenum. The pancreas also contains small cell islets that produce the hormone *insulin* (see Chapter 21).

The pancreatic juice is formed in two ways; there is a response to the ingestion of food due to a nervous reflex; but there is a more important response to the entry of acid gastric contents. This is a beautiful example of the work of chemical messengers, or *hormones*, in the body. When stomach contents enter the intestine, some of the acid extracts from the bowel wall a potent stimulating agent called *secretin*, which enters the bloodstream, is carried to the pancreas, and stimulates the secretion of pancreatic juice. The alkaline juice enters the duodenum amd neutralizes the acid influx, a feedback mechanism which automatically halts the mechanism for evoking pancreatic secretion. The next spurt of gastric contents starts a fresh cycle, and so on. Thus pancreatic secretion is exactly matched to the needs of digestion.

The pancreatic juice is as alkaline as gastric juice is acid. It contains three enzymes: *trypsin*, which furthers the digestion of proteins to amino acids; *amylase*, which hydrolyses carbohydrates to glucose or disaccharides; and *lipase*, which splits fats into fatty acids and glycerides. The pancreatic juice will not function properly without the presence of other secretions. Thus protein digestion by trypsin has to be initiated by the conversion of a precursor – *trypsinogen* – into trypsin proper by another enzyme – *enterokinase* which is liberated by the duodenal mucosa. Fat digestion is enormously helped by the bile, which emulsifies the globules to expose a far greater surface to the action of lipase.

Liver and bile

The liver is composed of myriads of lobules, each with a dual supply of ordinary arterial blood from the hepatic artery and of blood from the intestine containing products of digestion via branches of the portal vein. *Bile* is continuously secreted by the lobules to a total of some 850 ml a day. It is only required at intervals and is stored in the gall-bladder, where it is concentrated by reabsorption of water and released when required.

Bile is not only a digestive juice, but also a means of ridding the body of the breakdown products of haemoglobin. It is a slightly alkaline, viscid green fluid. The *bile pigments* are breakdown products of effete red cells, which disintegrate in the liver and spleen. These pigments, altered and mixed with undigested food, give the faeces their characteristic colour. In *jaundice*, due to obstruction of the bile duct which prevents the pigments entering the bowel, the faeces are pale and clay-coloured, while the dammed-up bile overflows into the circulation and stains the skin yellow. (Jaundice may also result from disease of the liver itself, as in infective hepatitis, or from blood destruction (*haemolysis*), and in these cases the faeces remain brown.)

The *bile salts* are the sodium compounds of certain complex acids and are essential in fat digestion. The bile contains no enzymes of its own, but its salts emulsify fat to facilitate its digestion by pancreatic lipase. The fats, split into glycerides and fatty acids, enter the bowel wall together with the salts; here, in the individual cells, globules of neutral fat are reformed and absorbed, while the salts return to the liver to be used again.

The muscular gall-bladder is normally full between meals. Entry of food into the duodenum relaxes the sphincter muscle guarding the opening of the common bile duct into the duodenum and the gall-bladder contracts and discharges bile.

Large intestine

The *large intestine*'s main function is the absorption of water from the faeces. Faeces enter the colon in a fluid state; here, they lose four-fifths of their water content, becoming solid in the lower (left) half of the colon. Much of the excreta consists of enormous numbers of the bacteria found in the large bowel. The stomach and upper small intestine are relatively sterile.

Thus, an abdominal wound that penetrates the large intestine may liberate bacteria into the peritoneal cavity, causing *peritonitis*.

The *faeces* are derived mainly from substances formed in the intestines. They contain 70% of water, some 15% of minerals such as calcium phosphate and iron salts, and much nitrogenous material, mainly from dead bacteria. An increased cellulose (fibre) content in the food increases the bulk of the faeces and stimulates bowel passage. The faeces also contain shreds of mucous membrane and are passed together with gaseous *flatus*, mostly methane.

Bleeding from the large bowel, if obvious, may be due to piles or a growth; *occult blood* found by chemical examination of the stools arises higher up, usually from a gastric or duodenal ulcer. Severe bleeding at a higher level produces black stools – *melaena*.

Movements in the digestive tract

Food is propelled onwards and downwards in the bowel by *peristalsis*. This is a rhythmical wave of contraction travelling down a segment or loop of bowel at about 2 cm (3/5 inch) a second, presenting a local ring-like spasm at any one point, preceded and succeeded by a wave of relaxation. Peristalsis is not dependent on central control, for it is brought about by a nerve plexus within the muscle coats of the intestine. It is however subject to modification by the central nervous system, as the effects of panic may show only too well.

In the small intestine there is another type of movement, *segmentation*, in which a length of bowel is demarcated into sausage-shaped segments which reform from time to time. This massages the chyme and propels it only slowly, giving time for digestion and absorption. Finally, there are *pendular* movements, waves of contraction passing backwards and forwards, churning the contained food.

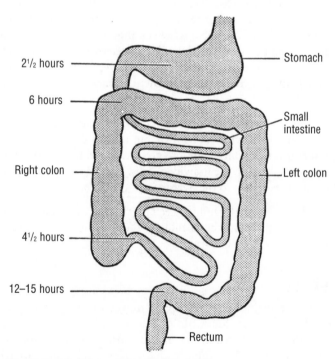

Figure 14.2 Rate of passage of food through the digestive tract

Regular peristalsis does not occur in the colon, where the problem is one of storage of the faeces until a new meal makes it necessary to move on its contents. Entry of food into the stomach causes a *gastrocolic reflex*, a mass contraction that transfers the faeces from left (descending) colon to rectum preliminary to *defaecation*. Defaecation consists of expulsion of the faeces outside the body, and the rectum normally remains empty until just before the act. Although this is reflex, it is normally under some conscious control by the external anal sphincter. The urge passes off if the call is not obeyed and the rectum relaxes. In chronic constipation the rectum always contains some faeces and loses its power of responding effectively to distension. During the act, the levator ani muscle (Figure 8.23) lifts the anal canal over the faecal mass, while evacuation is aided by voluntary contraction of the diaphragm and abdominal muscles to raise the intra-abdominal pressure. The external opening of the anal canal is normally closed by its two guarding circular muscle sphincters.

How long does it take for the food to reach different stages in its passage through the bowel? A meal of barium sulphate, which is opaque to X-rays, can be watched during its passage along the alimentary canal (Figure 14.2). The bulk of the stomach contents have passed on after three hours, though a small residue may persist for an hour or two. Food reaches the lower end of the small bowel and enters the colon after four to four and a half hours; fills the right side of the colon by six hours; and occupies the left colon, ready for expulsion into the rectum, at twelve to fifteen hours. These are average times and vary considerably; food may be retained in the large bowel for days.

Vomiting is a reflex action initiated by many different stimuli. These include:

▸ irritation of the stomach lining by food, alcohol or drugs

▸ touching the palate or pharynx

▸ distension or irritation of the stomach or bowel

▸ stimuli from the vestibular apparatus of the inner ear (Chapter 20) as in seasickness

▸ severe pain or emotional disturbance.

The vomiting centre is in the medulla of the hindbrain (Chapter 20) There is a violent contraction of the abdominal wall muscles, which expels the gastric contents through a relaxed oesophagus. Vomiting is usually accompanied by nausea, sweating, salivation, and an accelerated heart rate.

Test yourself

1 What is the name of the elastic flap of fibrocartilage that stops food going down the trachea on swallowing?
   a   Peristalsis
   b   Soft palate
   c   Epiglottis
   d   Larynx

2 What is food called when it enters the stomach?
   a   Bolus
   b   Chyme
   c   Fundus
   d   Fibre

3 What is the wave of contraction that propels food onwards?
   a   Peristalsis
   b   Segmentation
   c   Pendular
   d   Defaecation

4 Identify an enzyme present in gastric juice.
   a   Pepsin
   b   Bile
   c   Gastrin
   d   Lipase

5 What is the main blood vessel leading to the liver?
   a   Pulmonary artery
   b   Aorta
   c   Hepatic artery
   d   Vena cava

6 What is the pH of pancreatic juice?
   a   Acidic
   b   Neutral
   c   Alkaline
   d   Slightly alkaline

7 Where is bile stored?
   a   Pancreas
   b   Stomach
   c   Gall-bladder
   d   Liver

8 What is the main function of the large intestine?
   a   Absorption of carbohydrates
   b   Absorption of water
   c   Absorption of vitamins
   d   Absorption of minerals

9 What condition can occur if the bile duct is obstructed?

    a   Jaundice

    b   Peritonitis

    c   Appendicitis

    d   Melaena

10 Which part of the digestive tract has the slowest rate of passage of food?

    a   Stomach

    b   Small intestine

    c   Right colon

    d   Left colon

Absorption, utilization and storage of digested food

In this chapter you will learn about the absorption during digestion of:

▶ *water*

▶ *vitamins and minerals*

▶ *carbohydrates*

▶ *fats and proteins*

▶ *the metabolism of absorbed substances*

▶ *the role of the liver in metabolism.*

Absorption

Digestion transforms complex insoluble food substances into simpler soluble particles capable of entering the circulation. Their transference from the bowel to the blood or lymph is known as *absorption*. This is dependent on physical factors such as concentration and osmotic pressure, but most nutrients are *actively* transported and only water and alcohol cross the intestinal walls by simple diffusion.

Most nutrients are absorbed in the small intestine. The lining of the small intestine is not smooth but has a velvety pile of innumerable hair-like processes, or *villi*. These enormously increase the area available for absorption. Each villus (Figure 15.1) is richly supplied with blood capillaries, and is traversed by a central lymphatic channel or *lacteal* for carrying digested fat globules to the lymphatic trunks of the body. In the fasting state the villi are at rest, but during absorption there is a constant pumping motion which squeezes absorbed materials into the circulation.

In people with coeliac disease, the villi are defective and the bowel lining is smooth; this is associated with sensitivity to the gluten of cereal starches, and produces wasting and anaemia.

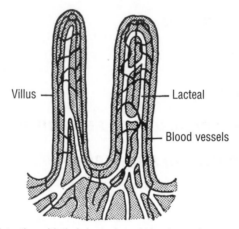

Figure 15.1 Villi of the small intestine with their lacteals and blood vessels

WATER

A lot of water enters the digestive tract daily. In addition to this *external load*, say 1500 ml, an *internal load* is contained in the digestive juices: saliva 1500 ml, gastric juice 3000 ml, bile 500 ml and pancreatic juice 200 ml – a total of 6700 ml. This is all absorbed from the bowel, most from the small intestine, the remainder in the large intestine, where the faeces become more solid. Most water enters the portal venous blood and only a little enters the lymphatics. The transference of water (and dissolved electrolytes) across the bowel wall into the vessels of the villi is not fully explicable in terms of omosis. There is a muscle slip acting as a mechanical pump and also a selective activity, or ionic pumping action, of the intestinal cells themselves. The absorbed water burdens the circulation and kidneys, which is why it is important to restrict intake for people with heart and kidney disorders. Excretion of excess water in the urine can be promoted by diuretic drugs.

VITAMINS AND MINERALS

The water-soluble vitamins are simple soluble molecules and are readily absorbed, although the absorption by the ileum of vitamin B12 depends on an *intrinsic factor* secreted by the stomach. The fat-soluble vitamins travel in the fat droplets across the bowel wall. Most of the calcium ingested is lost in the faeces, but some is absorbed to supply the needs of the skeleton, more so in growing children. This absorption is dependent on vitamin D. If there is insufficient vitamin D or the calcium intake is inadequate, rickets will develop in children or the equivalent bone softening – called osteomalacia or osteoporosis – in adults. Absorption of vitamin K, made by the resident bacteria, occurs in the large bowel. The absorption of iron is related to a person's requirements; excessive intake of iron leads to damaging deposits in the liver and tissues – *siderosis*.

CARBOHYDRATES

Carbohydrates are broken down by digestion into simple monosaccharides and disaccharides, and the latter are changed into the monosaccharide form by the cells of the bowel wall, followed by active transport into the portal blood. There are specific enzymes for the different sugars and, if these are absent, disaccharides may pass unchanged into the colon and cause diarrhoea; for example, some infants possess no lactase enzyme to digest lactose and cannot tolerate milk.

FAT (LIPIDS)

Very little fat escapes digestion and it is absorbed only from the small bowel. Fat leaves the stomach as large globules and is finely emulsified in the duodenum by the bile salts, and rendered easier to break down (hydrolysed) by pancreatic lipase into fatty acids and monoglycerides. These enter the cells of the bowel wall as a microemulsion and recombine within the villus cells into small globlules, or *chylomicrons*, of neutral triglycerides. These are then transferred to the lacteals and thence to the lymphatics. After a fatty meal these channels and the thoracic duct are distended by a milky fluid, the *chyle*. The thoracic duct empties into the great veins of the neck so that the blood serum may contain obvious fat a couple of hours after a meal. The process is accelerated by exercise.

PROTEINS

Under the influence of the pepsin of the gastric juice and the trypsin of the pancreatic secretion, proteins are broken down into their constituent amino acids and short-chain peptides. These are readily absorbed into the circulation and little of the dietary nitrogen is lost with the faeces.

Utilization (metabolism) of absorbed substances

PROTEIN METABOLISM

The place of proteins in the diet is dealt with in Chapter 13. We shall now consider their fate within the body, beginning with a look at their chemical composition. They are large-molecule, nitrogen-containing compounds with many differing functions. Some, like the keratin of hair and skin, are insoluble and inert; others fulfil the mechanical requirements of certain tissues – muscle and connective; some are soluble and used for transporting oxygen, minerals and hormones; and every enzyme is a protein. Humans are incapable of directly utilizing inorganic nitrogen and depend on ingesting the protein of other life forms, which is then degraded into its simpler amino acid constituents, to be recombined in human patterns.

The amino acids are the protein currency of the body – small, universally recognized, freely interchangeable and capable of being built up into the complex proteins of particular tissues.

All proteins can be hydrolysed, first to polypeptides and then to their consitutent amino acids, by boiling or treatment with a series of enzymes. *Simple proteins* so treated yield amino acids only. *Conjugated proteins* yield amino acids plus a linked prosthetic group that confers its specificity on the molecule. About 20 amino acids are obtainable from protein hydrolysis, and their basic chemical formula is:

$$R—CH—COOH$$
$$|$$
$$NH_2$$

where, in the simplest form, R = H (glycine).

In more complex forms, R may be a straight aliphatic chain or a cyclic radical. Amino acids can combine with both acids and bases and so act as neutralizing buffers. Soluble proteins can be isolated and identified by chromatography, which depends on their differing rates of diffusion in moist filter-paper into separate stainable protein 'spots', and also by electrophoresis, which depends on differential migration in filter-paper to one or other of a pair of electrodes.

The amino acids in a protein are linked by *peptide* bonds:

$$—CH—NH—$$

so that the basic structure of a protein molecule may be represented as:

$$R_1 \qquad\qquad R_2 \qquad\qquad R_3$$
$$| \qquad\qquad\quad | \qquad\qquad\quad |$$
$$—NH—CH—CO—NH—CH—CO—NH—CH—CO$$

where R represents specific amino acid radicals.

The molecular weights of proteins vary from a few thousand to over a million, i.e. the simplest protein molecule contains over a hundred amino acid residues and several thousand polypeptides. Though there are only some 20 amino acids, there are many more kinds of proteins because they differ in their constituent acids and how these are linked. They may be 'fingerprinted' by hydrolysing them to mixtures of peptides and then by electrophoresis, which yields a specific pattern for each (Figure 15.2).

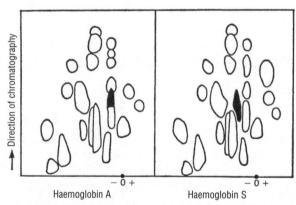

Figure 15.2 Fingerprints of tryptic hydrolysates of haemoglobins A and S. 0, point of application of hydrolysate; + –, direction of electric field

The peptides which differ in behaviour and composition are marked black; the one containing valine (in HbS) has moved slightly further towards the cathode (from *A Companion to Medical Studies*, Vol. 1)

The *shape* of the protein molecule may be compact and globular in the case of the soluble proteins, or rigid and elongated in the fibrous proteins, often insoluble. They are often folded in helical manner and identifiable by X-ray crystallography. Proteins may be *denatured* by heat, acids or alkalis, which loosen the internal bonds of the molecule.

PROTEINS IN NUTRITION

Energy is obtainable from all three basic foods, but only proteins serve the needs of growth and tissue repair.

There is great variation in the capacities of proteins to support life. Young rats fed on maize protein lose weight because it contains no lysine or tryptophane. If these are supplemented, growth resumes normally, i.e. these acids cannot be synthesized in the body, so they are known as *essential* amino acids and must be supplied in the diet. In humans the essential acids are: isoleucine, leucine, lysine, phenylalanine, tyrosine, histidine, methionine, cystine, threonine, tryptophane and valine. Hence the importance of a well-balanced diet. Living on a single staple risks causing protein deficiency. Note that the description 'essential' refers only to amino acids which are indispensable in the diet because they cannot be synthesized in the body; those that can are none the less vital.

The *biological value*, or net protein utilization, is:

$$\frac{\text{nitrogen retained in body}}{\text{effective dietary intake of nitrogen}}$$

a factor that can be calculated as:

$$\frac{\text{dietary N intake—faecal loss of N—loss of N in urine}}{\text{dietary N intake—faecal loss of N}}$$

The biological value of a reference protein such as milk is standardized as 100%, so the value is less for less utilized proteins. Any dietary constituent can be evaluated for its biological value and a *protein score* can be assigned to different foodstuffs.

The nitrogen required in the daily intake equals the daily loss plus that required for growth or other demands such as pregnancy or lactation. The nitrogen lost with the faeces and from the skin is largely independent of intake, but that in the urine varies with intake. The *obligatory nitrogen loss* is that lost daily in the faeces and urine and from the skin while on a protein-free diet. This obligatory loss is a measure of protein breakdown, or *catabolism*, and is increased by stresses such as infection, injury (including operations), pregnancy, lactation, fever and convalescence. Infants require more protein per kilogram of body weight due to their growth requirements and higher metabolic rate.

There is a relation between protein nutrition and calorie supply. Protein is used most efficiently when the bulk of energy needs is met by other foodstuffs, particularly carbohydrates. If this is inadequate, some protein must be diverted to supply energy and there is a risk of protein deficiency. If calorie needs are met by a diet poor in protein, it is impossible to eat enough to supply the intrinsic protein need. All the calorie needs of a growing child can be met by deproteinized milk; but protein deficiency will develop unless a supplement is given in the form of dried skimmed milk.

PROTEIN TRANSFORMATIONS IN METABOLISM

Proteins take part in many chemical processes; many different enzymes are involved and each protein is metabolized in a specific manner. The proteins of the tissues are not static but constantly receive fresh amino acid constituents from the blood and shed their old ones. The end-products of amino acid degradation are carbon dioxide and water, plus nitrogenous compounds such as urea and creatinine which are lost in the urine. Amino acids are present in the blood plasma and enter the interstitial (extracellular) fluid, to be taken up by cells, within which their concentration may be multiplied five- or tenfold. Little amino acid is normally excreted as such in the urine because of selective reabsorption back into the blood in the kidney tubules (Chapter 19). Amino acids are degraded by a process of *deamination*, an oxidative process in which the nitrogen is converted to ammonia. This is converted in the liver to urea, which is excreted by the kidneys. Most of the amino acid carbon is eventually lost as carbon dioxide but a significant amount of glucose is formed from the split protein. The sulphur of certain amino acids (cysteine and methionine) is excreted in the urine but some becomes linked to polysaccharides to form the ground substance of cartilage. Protein degradation is mainly by hydrolysis, liberating amino acids, some of which are used to replace effete intracellular proteins. Some idea of the length of life of protein molecules is obtained by tagging them with identifiable radio-isotopes; human serum albumin is replaced at the rate of 10% a day. The turnover of protein in the body as a whole is some 250 g a day, i.e. some 2.5% of the total body protein.

This leads to the concept of *nitrogen balance* between daily intake and excretion, mainly in the urine. In the steady state these are in equilibrium, but on a protein-free diet the balance becomes negative and urinary output of nitrogen falls to the basal obligatory loss. The weight and protein content of the liver and other organs fall and cell division and replacement slow down. Return to a normal diet reverses these changes; there is a positive nitrogen balance until equilibrium is restored.

FAT (LIPID) METABOLISM

Fats are compact, readily mobilized, rich stores of energy. They are insoluble in water. The main groups are:

1 fatty acids

2 triglycerides and waxes

3 phospholipids

4 cholesterol and its derivatives

5 vitamins A, D, E and K.

The *fatty acids* are unbranched aliphatic chains with a terminal carboxyl (COOH) group. The simplest form is acetic acid, CH_3COOH, and the type formula of more complex varieties is:

$$CH_3-(CH_2)_n-CH_2COOH$$

Fatty acids may be *saturated* or *unsaturated*, the latter having the capacity to accept more hydrogen atoms; commercially, hydrogenation is used to convert unsaturated vegetable oils to solid fats. Unsaturated fatty acids are considered valuable in preventing arterial disease.

The *triglycerides* comprise the bulk of dietary fat; their basic model is an ester of glycerol with three fatty acid chains. All fats are triglycerides, differing only in the nature and amount of the fatty acids esterified to the glycerol base. In terms of the number of carbon atoms in the fatty acid chain, human fat consists mainly of C_{16} and C_{18} acids, saturated or unsaturated.

Fish oils have a high content of polyunsaturated fatty acids and are liquid and very healthy in preventing atherosclerosis. Saturated lard and suet are solid at room temperature.

The *phospholipids* contain phosphorus and sometimes nitrogen; they are important components of cell membranes, plasma and the nervous system.

Cholesterol is a very important compound with a basic ring structure:

Some comes in the diet; more is synthesized in the body, notably in the liver. It is an example of a class of lipids known as *steroids*. These are important compounds and include the hormones of the adrenal and sex glands, vitamin D and the bile acids.

THE FATE OF ABSORBED FAT

The liver plays a cardinal role in fat metabolism; it normally contains some 5% of fat, but more may appear in diseases such as fatty degeneration, starvation and certain forms of chemical poisoning.

About 100 g of fat is absorbed daily, most as triglycerides. They enter the duodenum as a crude emulsion, are finely dispersed by the bile salts, and split by pancreatic lipase to fatty acids, monoglycerides and some glycerol. These are not water-soluble and their transfer across the bowel lining is facilitated by bile salts, which form them into tiny aggregates, *micellae* or *chylomicrons*, which enter the villi. Within the villus cells they recombine, though not into the original triglycerides. They enter the lymph vessels in small particles or chylomicrons, which enter the central lacteal of the villus and travel up the thoracic duct.

Lipids are transported in the plasma as *lipoproteins*, large molecules of triglycerides embedded in cholesterol, phospholipids and protein. These convey the triglycerides to the tissues, where they are separated for local use, while the conveyor system shuttles back to the liver to pick up more triglycerides. The *serum lipids* consist of:

1 *cholesterol*, in amounts of 150–250 mg/100 ml in youth, rising to 300 mg or more in later life

2 *phospholipids* in amounts of 150–300 mg

3 *triglycerides* in amounts of 50–300 mg in the fasting state, reaching much higher levels after a fatty meal.

The cholesterol goes to the liver, where it is converted into bile salts and excreted into the bowel. However, in atherosclerosis, cholesterol is deposited in the lining of the arteries and may obstruct them.

In the tissues the arriving lipoproteins are cleared from the plasma and the triglycerides are split by lipase. The fatty acids produced enter the fatty tissue and are reconverted to triglycerides or they enter muscle to supply energy by oxidation. The freed glycerol returns to the liver. Thus part of the transported plasma fatty acids is stored in the body fat and part is burnt to supply the energy for muscle contraction. During exertion the fatty acid content of the plasma rises and more is taken up by muscle. This mechanism is less important if there is ample free glucose available. In brief exercise the initial source of muscle energy is the glycogen stored in the muscle cells, which is split to yield glucose. But after some minutes the muscles begin to take up free fatty acids from the blood.

The energy store represented by the triglyceride content of adipose tissue is available when required, the triglycerides being split back to glycerol and fatty acids which enter the blood, from which the acids are taken up by the organs and oxidized to yield energy.

The carbohydrates of the diet may also contribute to the fat store. The glucose molecule is broken down and the carbon atoms reassembled in long fatty acid chains.

Cholesterol is synthesized in the tissues, and that of the blood is partly free and partly esterified. The phospholipids are essential to the integrity of cell membranes, which control passage of various small molecules, nutrients, wastes and water.

High-density lipoproteins (HDLs) contain more protein and *low-density lipoproteins* (LDLs) more fat. As the HDLs transport cholesterol for breakdown, high levels (above 60 mg/100 ml in the blood) are considered protective against cardiovascular disease. High levels of LDLs (160 or more) are considered sinister as they promote the deposition of cholesterol in the arterial walls.

FATTY ACIDS AND CARBOHYDRATES

Fatty acids and carbohydrates are broken down in the tissues to yield carbon dioxide, water and energy. In the resting state most of the energy requirement is provided by the fatty acids, less from glucose, and even less from amino acids, ketones and glycerol. But this changes with eating and exercise. After a carbohydrate meal the respiratory quotient moves nearer to 1, i.e. glucose rather than fat is being used. During strenuous exercise the main initial source of energy is from the glycogen store of muscle, but if the exertion continues the muscles begin to use free fatty acids from the blood and the respiratory quotient falls. During fasting the liver glycogen is depleted, the blood glucose falls, less is used in the tissues, and the metabolism turns over to fat. However, the brain always prefers glucose for energy transformations.

KETOSIS

If insufficient glucose is available because of fasting or prolonged exertion, energy is gained predominantly by the mobilization of fat stores. Metabolism of fats yields *ketone bodies*, such as acetoacetate, and this lowers the pH of the blood – a condition of *acidosis*. Ketone bodies also appear in the urine. Similar changes may occur in uncontrolled diabetes. A diet where the fat/carbohydrate ratio is too high also leads to ketosis, since fats burns best in the fire of carbohydrates.

FATTY (ADIPOSE) TISSUE

The basic fat is widely distributed and makes up some 10% of body weight. About half is under the skin; the rest is mostly in great sheets in the abdominal cavity (the omenta, Chapter 8), or as packing around the internal organs, a little in the intermuscular planes. There is none in the cranial cavity. A healthy woman has about twice as much fat as a man, but the amount in any one person depends on constitutional and nutritional factors. Its distribution is affected by the sex hormones; there is more fat stored round the hips and thighs in women and more fat stored in the upper trunk in men.

Fat insulates the body against heat and cold and buffers the internal organs. Fat is also an electrical insulator around nerve fibres. Microscopically, fat is divided into lobules by fibrous partitions. The fat cells are connective tissue cells distended and deformed by a central fat globule, which pushes the nucleus to one side.

Chemically, fat stores fatty acids in the form of triglycerides as a source of energy when required. It consists of some 85% actual fat (mostly neutral triglycerides, though there are also some free fatty acids, glycerol, cholesterol and phospholipids), 12% of water and 3% of protein. But adipose tissue is not a passive storehouse; it is an entrepôt where continuously arriving new material is broken down and resynthesized to pass on to the tissues. Fat can be formed even on a fat-free diet from glucose in the blood, synthesis occurring in the liver and other tissues.

Fat is added to the reserve when the food exceeds energy needs. Different animals have characteristic fats; human fat is liquid at body temperature, whereas mutton fat is solid. This indicates the body's ability to modify fat in the diet.

Fear, cold and other stresses mobilize fatty acids, and this response is mediated partly by adrenal hormones and partly by the autonomic nervous system.

CARBOHYDRATE METABOLISM

Glucose is manufactured from carbon dioxide and water by photosynthesis in green plants. It is rarely found as such in nature but occurs in polymer form as starch, which is a polymer in cereals that humans can digest.

The basic building block of carbohydrates is CH_2O and their general formula is $(CH_2O)n$. The commonest carbohydrates are *monosaccharides*, where n = 5 or 6, *pentoses* and *hexoses*. These are soluble and sweet. *Disaccharides* such as maltose are made by linkage of two monosaccharides. Ascorbic acid (vitamin C) is a hexose derivative.

Disaccharides exist as *stereoisomers*, i.e. the same compound has different asymmetric spatial arrangements of atoms in the molecule. The number of possible isomers may be large – 16 for a hexose. Although the chemical analysis of isomers is identical, they differ in their physical and biological properties. However, the body can use only one isomer of any one compound; the others, though chemically identical, have little or no physiological activity. Thus l-ascorbic acid is active against scurvy whereas d-ascorbic acid is not; and l-thyroxine stimulates general metabolism while d-thyroxine is only a third as effective. When such a compound is made in the laboratory, it consists of a *racemic mixture* of equal proportions of both isomers, which is optically inactive since the rotations of the constituents cancel out, and it has only half the biological activity of the natural isomer.

In the metabolism of glucose, a basic reaction is with adenosine triphosphate (ATP) (a nitrogen-containing purine body which yields glucose-6-phosphate and adenosine diphosphate (ADP)). This is the great work-horse of metabolism and the reaction is catalysed by an enzyme, hexokinase. Glucose-6-phosphate is the basic carbohydrate currency of the body, just as the amino acids are the basic protein currency. It is the starting-point for glucose/carbon metabolism, with many pathways and intermediate products, in which organic phosphates and phosphatase enzymes are prominent. The basic final degradation of glucose to yield energy is to pyruvate (pyruvic acid is CH3(CO)COOH), which is split in the tissue cells to carbon dioxide and water.

Glycogen is found everywhere but mostly in liver and muscle. It is a polymer of glucose with a molecular weight of several million.

Lactose is the disaccharide of milk, synthesized from the monosaccharides glucose and galactose only in the breast. In the infant it is hydrolysed in the bowel wall to glucose and galactose, and the latter is converted to glucose. If the last sequence is impossible for lack of the appropriate enzyme, galactose accumulates and may cause stunting or mental defect; the only cure is a galactose-free diet.

The liver is the only organ that can form free glucose. Blood sugar cannot be permitted to fall below a minimum threshold without grave consequences. This is because there is a continual demand for glucose from the blood by the various organs, particularly the brain. Therefore, even in starvation or on a carbohydrate-free diet, the liver must manufacture glucose for as long as possible.

As with protein metabolism, the sugar of the blood is like a reservoir that is simultaneously being emptied and refilled. There is a constant intake from food and liver glycogen and a constant output into the tissues, where some is burnt for energy and some stored as a glycogen reserve. The blood sugar is thus in equilibrium with that of the tissues and a very efficient mechanism keeps the concentration fairly constant. It rises somewhat after a meal but is back to normal within two hours. This regulation is so effective that it remains much the same in starvation as after a heavy meal. Only exceptionally does the blood sugar level rise so much that some sugar spills over into the urine.

The regulator of blood sugar levels is *insulin*. Insulin is an internal secretion or hormone of the pancreas. Insulin is essential to the utilization of sugar by the tissues and is normally formed by the pancreas in amounts just sufficient to keep the blood level down to normal by enabling the tissues to use added sugar. Without insulin, the tissues cannot properly burn glucose, sugar is dammed up in the blood and enters the urine. The loss of the energy source produces relative starvation; also fat metabolism is interefered with and acidosis results. This is the condition of *diabetes*, due to deficient insulin secretion. The treatment consists of giving insulin artificially by injection, and of balancing the insulin given and the sugar in the food so that the blood level is kept steady. In minor cases of *maturity-onset diabetes*, management may be possible by reducing sugar intake and by drugs. Excess of insulin abnormally lowers the blood sugar, possibly with serious consequences – acute anxiety, hunger, pallor, sweating, convulsions, coma and even death from *hypoglycaemia*. Slight overproduction of insulin in normal persons may cause recurrent restlessness or even mental disturbance, rapidly alleviated by eating sugar or sweets.

The liver is the chief store of glycogen and constantly converts into glucose and vice versa; it also forms glycogen from protein and fat. Just as insulin regulates sugar outflow by enabling the tisues to burn it, so the liver regulates inflow by adding to its glycogen reserves. This liver store is a reserve against starvation and quickly converted into soluble glucose and passed into the circulation. Of all the body's reserves, this is most rapidly available and easily used. In its mobilization *adrenaline*, the internal secretion of the adrenal glands, acts as a potent messenger to the liver.

The muscles are the other main stores of glycogen, to supply their demands in contraction. They convert it, not into glucose, but *lactic acid*, which enters the blood and is reconverted into glycogen by the liver or the muscles. More is stored in the muscles than in the liver, but is kept for exertion and is not readily available for general use.

The overall fate, distribution and recycling of carbohydrates within the body is indicated in the following diagram:

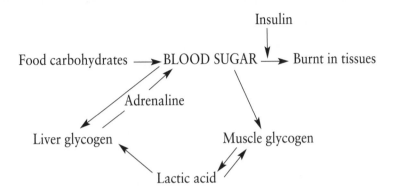

THE LIVER IN METABOLISM

The liver is of cardinal importance in metabolism. It plays an important part in the synthesis, storage, degradation and interconversion of protein, fat and carbohydrates. Bile formation, the synthesis of urea from ammonia radicals split off from proteins, the metabolism of drugs, hormones and toxins: all are the exclusive province of this organ. If the liver ceases to function normally, serious illness – even death – may follow. Yet it has a huge functional reserve and great powers of regeneration; even small portions can hypertrophy and maintain life, and there may be no obvious disturbance even when when it is extensively diseased.

Because the liver is so important in the synthesis of proteins, the interconversion of amino acids, and the production of urea as an end-product of protein metabolism, in liver failure the amino acid content of the blood rises and there is a fall in urinary urea and blood urea. Protein is formed in the liver partly from dietary amino acids and partly from amino acids derived from tissue breakdown. Protein is broken down by deamination, the removal of amino acids, and the potentially toxic ammonia so formed is converted into non-toxic urea. Other organs, notably the kidney, also form ammonia, and this is brought to the liver for conversion.

The liver plays a special part in the formation of the *plasma proteins*. These include:

▶ *albumin*, the most abundant protein in the blood plasma

▶ *alpha-* and *beta-globulins* in smaller amounts

▶ proteins concerned with blood clotting

▶ certain enzymes.

The plasma proteins have a short life and are constantly renewed. Their analysis may be effected by electrophoresis, which gives a definite pattern (Figure 15.3), which changes in disease. If the diet is poor in protein, the plasma proteins fall and their osmotic action of retaining water is reduced, so that in famine conditions fluid seeps out of the blood to cause oedema of the soft parts and subcutaneous tissues. The starving person looks bloated.

We have already seen the liver's function in maintaining the blood glucose level and the part played by hormones – insulin and adrenaline – in this mechanism at rest and execise. The liver normally stores about 100 g of glycogen, immediately available to maintain blood glucose. But it also actively manufactures glucose from glycerol freed by fat breakdown, from monosaccharides, and from the breakdown products of some amino acids. This *gluconeogenesis* is promoted by certain hormones: the adrenal hormone *cortisone*, *thyroxine*, and a pancreatic hormone, *glucagon*.

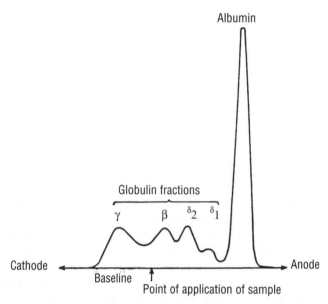

Figure 15.3 Densitometer tracing of an electrophoretic separation of normal serum proteins (from *A Companion to Medical Studies*, Vol. 1)

The liver both synthesizes lipids arriving from fat depots and the food and breaks them down, the long-chain fatty acids being oxidized to water and carbon dioxide. If this breakdown is impaired, ketone bdies are formed – acetocetic acid, hydroxybutyric acid and acetone – causing *ketosis* or *acidosis*.

Ketosis is an index that metabolism is predominantly of fat, as in starvation or prolonged fever, after long exertion, or in uncontrolled diabetes.

Cholesterol is synthesized in the liver (and elsewhere) and excreted in the bile to be reabsorbed from the intestine. The liver is also important in the formation of *lipoproteins*. Nine-tenths of ingested alcohol is oxidized in the liver, where it may cause cirrhosis if excess is consumed. The bile salts are formed in the liver from cholesterol.

Of the *bile pigments*, the most important, *bilirubin*, is formed by breakdown of the haemoglobin of the red blood corpuscles. This reaches the liver and is modified and excreted in the bile to enter the bowel. There, bacteria convert it into *stercobilin*; some of this colours the faeces, some returns to the liver and circulates repeatedly between bowel and liver or is excreted by the kidneys as *urobilin*, the yellow pigment of urine. *Jaundice* occurs when the normal level of bilirubin in the plasma is exceeded. It may be due to excessive production of bilirubin, due to accumulation in the plasma, disease of the liver, or obstruction of bile outflow from the liver. The first type occurs when red cells are destroyed in unusually large numbers, as in some anaemias or malaria. Liver disease impairs bilirubin transport. Obstruction to bile outflow, as by a stone or cancer in the bile ducts, dams back the pigment. In most forms of jaundice, bilirubin appears in the urine. Newborn infants are often slightly jaundiced for a few days after birth, and this is normal.

The liver also metabolizes certain hormones, as well as many drugs and toxic agents. If it cannot, liver failure may result, one feaure of which is fatty degeneration. The liver is also the chief depot of the fat-soluble vitamins A, D, E and K, and holds enough of the first two to prevent deficiency disease for a long time, even if they are absent in the diet. Vitamin K is used for the synthesis of clotting factors. Water-soluble vitamins are also stored, and the liver's enzyme systems can convert dietary tryptophane into nicotinic acid (Chapter 13).

There are various *tests of liver function*. These include measurement of plasma bilirubin or albumin, the former being raised and the latter reduced in liver insufficiency. A galactose tolerance test designed to assess the organ's ability to convert galactose to glucose and a test for escaped liver enzymes in the urine are also available.

Test yourself

1 What are the hair-like processes in the small intestine that increase its surface area?
   a   Alveoli
   b   Villi
   c   Lacteal
   d   Ileum

2 Which digestive juice contributes the most to the internal load of water?
   a   Saliva
   b   Bile
   c   Gastric juice
   d   Pancreatic juice

3 Which drugs can be used to excrete excess water in urine?
   a   Beta blockers
   b   Anabolic steroids
   c   Diuretics
   d   Amphetamines

4 Which vitamin is required to ensure calcium is absorbed by the body?
   a   Vitamin A
   b   Vitamin B
   c   Vitamin C
   d   Vitamin D

5 What form of carbohydrates are absorbed by the body?
   a   Monosaccharides
   b   Disaccharides
   c   Polysaccharides
   d   Peptides

6 What are enzymes made of?
   a   Carbohydrate
   b   Fat
   c   Protein
   d   Minerals

7 How are amino acids linked to form proteins?
   a   Saccharine bonds
   b   Peptide bonds
   c   Hydrogen bonds
   d   Conjugated bond

8 Which of the following increase catabolism?
   a   Injury
   b   Excessive vitamin intake
   c   Excessive food intake
   d   Sedentary lifestyle

9 What condition can fish oils help to prevent?
   a Atherosclerosis
   b Anaemia
   c Rickets
   d Osteoporosis

10 Which of the following is a bile pigment?
   a Haemoglobin
   b Bilirubin
   c Stercobilin
   d Urobilin

16

Blood, lymph and the reticuloendothelial (immune) system

In this chapter you will learn about:

▶ *red and white blood cells*

▶ *blood clotting, groups, volume and pH*

▶ *lymph and the reticuloendothelial system*

▶ *the lymphatic system*

▶ *the thymus*

▶ *immune-deficiency syndromes.*

Blood

The blood is a complex fluid pervading all parts of the body – except cartilage and the cornea – that conveys food and oxygen to the tissues and carries wastes away. It varies, according to its oxygenation, between the purple of the venous blood and the bright red in the arteries,. Microscopic examination shows myriads of *corpuscles* of different kinds suspended in the fluid component, or *plasma*. Corpuscles and plasma occupy an approximately equal volume and can be separated by standing or centrifuging, which settles the solids to the bottom. The blood cells include:

▶ red cells, or *erythrocytes*

▶ white cells, or *leucocytes*

▶ *platelets*, or *thrombocytes*.

In centrifuged blood the deposit consists of red cells; above this is a thin layer of white cells and platelets, the *buffy coat*; and above this is the clear yellowish *plasma*.

When blood is allowed to clot, the remaining fluid part is called the *serum*; this is almost, but not quite, identical with the plasma.

The plasma is a clear, pale yellow fluid containing some 10% of solids – mostly protein – together with salts, particularly sodium chloride – also sodium bicarbonate, phosphates and potassium, and metabolites from the different foodstuffs – glucose, urea, amino acids, fatty acids, etc. The *proteins* of the plasma are coagulable by heat or acid. One protein, fibrinogen, is concerned exclusively with clotting. The albumin and globulin are responsible for the considerable plasma osmotic pressure.

Blood cells

RED CORPUSCLES

The *red corpuscles* are by far the most numerous, some five and a half million per cubic millimetre. They are not true cells since they have lost their nuclei – biconcave discs, narrow-waisted in profile, which are envelopes for the respiratory pigment haemoglobin. This splitting of haemoglobin into innumerable little packets vastly increases the surface area for oxygen interchange. The total area of the red cells is 1000 to 2000 times that of the body. They are sensitive to changes in the osmotic pressure of the plasma. Increased pressure shrivels them by withdrawing water, while they swell up and burst in a weaker solution – their pigment is freed and the blood is said to be *haemolysed*.

Before birth the red cells are formed in the bone marrow, liver, spleen and lymph nodes. After birth the marrow is their only source; in early life the bones are full of red marrow, but later this retreats to the bone ends, leaving the shafts occupied by yellow fatty marrow. The survival time of the adult red cell is some 120 days, after which it fragments and is absorbed. They develop from specialized cells which lose their nuclei; but intensive demands, as in response to haemorrhage or anaemia, results in the temporary appearance of nucleated forms.

If blood treated with an anticoagulant is allowed to stand, the red cells sediment slowly into clumps and fall to the bottom. The *rate of sedimentation*, or ESR, is influenced by the composition of the plasma proteins; it is often increased in inflammation or disease and is an index of the activity of the process over time.

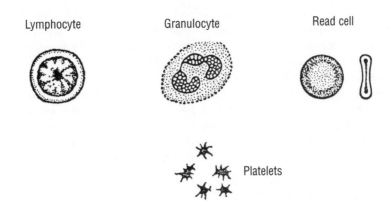

Lymphocyte   Granulocyte   Read cell

Platelets

Figure 16.1 Different types of blood corpuscles

Haemoglobin consists of a protein – *globin* – linked to a chemical grouping – *haem* – containing iron, which is responsible for its red colour. Its value is the ease with which it combines loosely with oxygen in the lungs to form bright red *oxyhaemoglobin* and with which this oxygen is given up to the tissues, leaving purple reduced *deoxyhaemoglobin* (80% of the carbon dioxide is dissolved in the plasma; the rest is carried to the lungs by the red cells).

Many factors are essential for the maturation of red cells from stem cells in the red marrow and to provide the right amount of haemoglobin, and a lack of any one may cause anaemia. Amino acids are required for the synthesis of globin and iron is essential because of its central position in the haemoglobin molecule. Women need more iron than men because of blood loss in menstruation, and the growing child proportionately more than the adult. So iron-deficiency anaemia is common in women and children but uncommon in men unless they are losing blood from a stomach ulcer or a blood-destroying parasite. We mentioned the importance of vitamin B12 and folic acid and the gastric intrinsic factor essential to the utilization of the B12 vitamin (Chapter 13); lack of any of these is the cause of *pernicious anaemia*. Other essential factors include thyroid hormone, copper and vitamin C, and a kidney hormone called *erythropoietin*.

In health a red cell of normal shape and size is termed *normocytic*, and, if it has a normal complement of haemoglobin, *normochromic*. Smaller or larger cells are *microcytic* or *macrocytic*; those with reduced haemglobin are *hypochromic*. The normal haemoglobin content of blood is 13.5–18 g/100 ml in men and 11.5–16.5 g/100 ml in women. Anaemia exists when the level is reduced because there are fewer red cells, or their haemoglobin content is reduced, or both. These factors can be expressed numerically by relating haemoglobin and red cell count to *packed cell volume* (PCV), the proportion of blood volume occupied by packed red cells after centrifuging, some 40–55%. This gives the mean cell volume (MCV) and the *mean cell haemoglobin concentration* (MCHC):

$$\text{MCV} = \frac{\text{PCV\%}}{\text{Red cell count} \times 10^6 \text{ mm}^3} \times 10$$

$$\text{MCHC} = \frac{\text{Haemoglobin (g/100 ml)}}{\text{PCV\%}} \times 100$$

Ordinary anaemia is due to lack of iron, where both the MCV and MCHC are reduced, i.e. the red cells are smaller and paler than usual and the haemoglobin may fall by half: microcytic or hypochromic anaemia. In pernicious anaemia, however, the red cells, though fewer, are larger than normal and have their usual complement of haemglobin: macrocytic anaemia. Anaemia causes pallor of the skin and mucous membranes and breathlessness, and easy fatiguability due to reduced oxygen-carrying power of the blood. The blood can adjust to even severe anaemia if its onset is gradual. It may be possible for a person to walk normally with a third of the normal haemoglobin if the cause is chronic, whereas a rapid fall to a much higher level may cause breathlessness. The symptoms of listlessness and inertia can be due to severe chronic anaemia produced by parasitic disease or dietary deficiencies.

In *sickle-cell anaemia* there is an abnormal haemoglobin S, the cells are crescentic and tend to clump and obstruct small vessels in sickle-cell crises. This is a genetic condition in African populations and acts as a defence against malaria by denying the parasites potassium.

WHITE CELLS (LEUCOCYTES)

There are three varieties:

▶ polymorphonuclear granulocytes

▶ lymphocytes

▶ monocytes.

The total white blood cell count is 4000–10,000/mm³. If this is markedly increased, as in infective disease, this is called a *leucocytosis*; if decreased, as in toxic bone marrow failure, it is a *leucopenia*. Increase due to a malignant cell multiplication is a *leukaemia*.

The *polymorphs* (granulocytes) have a segmented nucleus with a granular cytoplasm and form two-thirds of all the white cells; 90% are neutrophils and a few are eosinophils or basophils (referring to their affinity or not for acid or basic dyes).

The *lymphocytes* have a large central basophil nucleus occupying most of the cell; nearly a third of white cells are lymphocytes. The *monocyte* is the largest and scarcest: some 3–4%. It has abundant cytoplasm and an indented or lobed nucleus.

The polymorphs are produced only in the red marrow and there are large reserves. Their life-span is 1–2 weeks. They are lost in defence operations (see below) and are also shed into the body cavities.

The lymphocytes are formed in lymph nodes, the tonsils, spleen and lymphoid patches in the small bowel, and in the bone marrow. Most are actually outside the blood, as a large reserve in lymphoid tissues. There are two types, large and small. They are not phagocytic but are essential parts of the immune system, combating infections in two ways. The *T lymphocytes* act directly against noxious organisms and damaged cells (and also tumour cells). The *B lymphocytes* form *plasma cells* which produce antibodies (immunoglobulins) released into the blood.

The leucocytes are essential in fighting infection and inflammation. The mobile polymorphs destroy and remove foreign matter and bacteria, while the lymphocytes produce antibodies. The polymorphs also have the power of *phagocytosis*, or swallowing foreign substances. At the site of injury or infection they ooze through the capillary walls and assemble to overcome the microorganisms and remove waste materials, including dead and damaged cells. If they are killed, their dead bodies form part of the pus. The monocytes and neutrophil polymorphs are the most mobile in the tissues and are attracted to sites of damage by locally released chemical agents. They flow round particles and engulf them in a streaming amoeboid movement. The other two types of polymorphs are less active. The eosinophils combat parasitic worms and allergens. The basophils form histamine, a vasodilator which may be overactive in sensitization reactions.

The *platelets* are tiny refractile bodies lacking a nucleus and number $150,000–350,00/mm^3$. They have the radiating processes of an asterisk in shape. They are not true cells but fragments of large bone marrow cells shed into the circulation and assist in the process of clotting.

CLOTTING

Blood clots a few minutes after being shed. The clot is a meshwork of a substance called *fibrin*, derived from the fibrinogen protein of the plasma with blood cells trapped inside. This soon contracts, squeezing out a straw-coloured fluid, or *serum*, which is what remains of plasma after the loss of its fibrinogen. Serum and plasma are largely, but not entirely, identical.

Blood (plasma + corpuscles) ⟶ clot ⟶ (fibrin + corpuscles) + serum

Clotting is a complex process requiring the presence of calcium salts. The fibrinogen-fibrin conversion is activated by an enzyme, *thrombin*, which is not present in normal blood but formed from an inactive precursor, *prothrombin*.

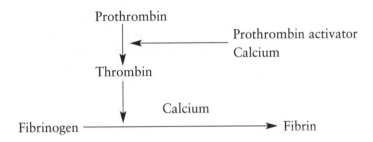

The process is complicated for there are many accessory factors. Clotting may take place in injured blood vessels or in shed blood in the tissues, or even within arteries and veins themselves if these are diseased. When a vessel is divided, the injured ends retract, platelets are attracted to the damaged endothelium and aggregate to seal the cut ends. After this initial phase of platelet plugging, the secondary phase of clotting occurs. As the process must not be allowed to spread back through the vascular tree, there is a complementary process of *fibrinolysis*, in which the fibrin is dissolved except where needed. Clotting is facilitated by substances derived from injured tissues, and if blood is carefully collected in a perfectly clean vessel it remains fluid for a considerable time. Other clotting agents include vitamin K, which is stored in the liver. Blood may fail to clot normally in certain conditions. An essential factor is constitutionally absent in haemophilia, which is a sex-linked genetic defect transmitted to their male descendants by women who do not express the condition themselves. Certain chemical agents delay or abolish coagulation. One of these is *heparin*, isolated from the liver; another is *dicoumarol*, analogous to the 'warfarin' used to poison rats. They are useful in surgery to prevent postoperative vein thrombosis and in the repair of injured blood vessels.

Nugget
Some snakes and leeches inject anticoagulants into their victims.

BLOOD GROUPS

Human tissues and fluids are sensitive to foreign proteins and react against them. In solution, this causes their precipitation. Human plasma mixed with animal blood causes the latter's cells to *agglutinate* together in sticky clumps. If animal blood is transfused these obstruct the recipient's blood vessels and their debris blocks the kidneys.

This sensitivity may also exist between one individual and another if they belong to different blood groups. Within the same group transfusion is safe; outside this only certain combinations can be used. Blood groups are inherited, with specific antigens (see below) on the surface of their red cells. There are four main groups: O, A, B and AB. Their compatibility is shown in the table. Note that it may be possible for a member of one group to donate blood to another, though fatal to receive blood from the latter.

Group	Can give blood to	Can receive blood from
O	O, A, B, AB	O
A	A, AB	O, A
B	B, AB	O, B
AB	AB	O, A, B, AB

Blood groups are inherited and were useful in cases of disputed paternity before DNA testing came into use. The distribution of blood groups differs in different parts of the world.

The red cells of some people possess a component analogous to one in the blood of the rhesus monkey, the *rhesus factor*. Most people (some 85%) have this factor; they are Rh positive and can have children normally. But if an Rh negative female becomes pregnant by an Rh positive male, the Rh factor inherited by the foetus from its father excites an antagonistic reaction in the mother. Agents formed in her blood destroy the red cells of the child, which may be stillborn or with serious anaemia or jaundice, or cerebral palsy because of brain haemorrhage. The disease is rare in a first pregnancy, but increases with successive pregnancies and may require exsanguination-transfusion of the newborn.

BLOOD VOLUME AND pH

The total volume of blood is 4–5 litres, varying with the size of the individual. After severe haemorrhage the fluid loss is made good by entrance of water into the circulation from the reserves of intracellular fluid; there is dilution and a lower red cell count. The reverse occurs in severe shock or dehydration; in these instances, plasma leaks into the tissues, leaving a viscous over-concentration of corpuscles in the vessels.

The blood is slightly alkaline, with a pH of 7.4. If we try to neutralize this with acid, more is required than would be expected. This is because there is an *alkali reserve* of sodium bicarbonate which, together with the plasma proteins, acts as a buffer, resisting any change in hydrogen ion concentration. When this reserve is greatly reduced, as in ketosis, true acidosis develops, but this is very rare.

Lymph

Lymph is a pale yellow fluid, similar in composition to interstitial (extracellular) fluid, which clots on standing and contains lymphocytes. The lymphatic system of the bowel plays the main part in absorption of fat through the lacteal vessels of the intestinal villi, so that lymph in the thoracic duct after a meal is milky and opalescent wth fat globules. (For details of the lymphatic system, see below).

Reticuloendothelial (immune) system

This is not localized, but is a tissue widely dispersed throughout other tissues and organs. It consists (Figure 16.2) of a meshwork of fine fibres supporting large irregular cells and permeated by channels, or *sinusoids*, lined by an endothelium composed of these cells.

These cells are phagocytic and some wander in the tissues as scavengers. The monocytes of the blood form part of this system, and phagocytosis is an essential defence mechanism in fighting invasion by bacteria or parasites. The body is exposed to infective attack from three sources:

1 *viruses*, like those of poliomyelitis and chickenpox, which are too small to be seen with the optical microscope and traverse very fine filters

2 *bacteria*, microscopically visible organisms like those of plague and typhoid

3 larger parasites, which are single cells but still microscopic, like the *protozoa* of malaria and other diseases.

The primary defence is the outer coverings and internal linings of the body, the normally intact 'epithelial envelope'. The skin is a physical obstacle. The mucous membranes are more sensitive, but even here organisms become entangled in mucus or are wafted away by ciliary action in some regions, as in the airways. The secretions of certain glands are bactericidal, as with the tears and saliva. An intact epithelium is normally invasion-proof, but the gonococcus and spirochaete of gonorrhea and syphilis and the AIDS virus can traverse the intact mucous linings of the genital tract.

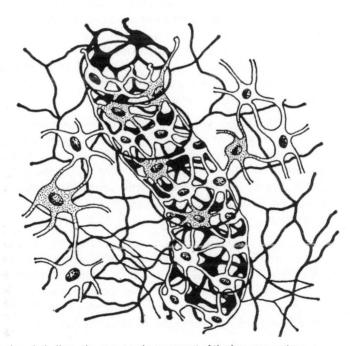

Figure 16.2 Reticuloendothelium, the structural component of the immune system

Note the network of reticulin, the reticulum cells and the endothelial-like lining they form on the reticulin framework of the fluid channels or sinusoids (from *A Companion to Medical Studies*, Vol. 1)

Once organisms have entered the tissues, the first defence is their being attacked by the phagocytes of the immune system, which wander to the affected area and devour debris and organisms. Minor processes can be managed by the phagocytes locally available, but it is usually necessary to bring up reinforcements by increasing the local blood supply to provide many more phagocytes. All the capillaries dilate widely and the granulocytes make their way through the capillary walls to the site of infection, where they accumulate. When the infection is overcome and the debris removed, repairs are carried out by an overgrowth of young fibrous (granulation) tissue in the depths of the zone, and an ingrowth of surface epithelium if there has been an open wound.

Although the reactions to injury and infection are similar, the latter requires many more phagocytes. In an infective focus still unresolved after 48 hours, *pus* (a fluid collection of dead phagocytes and bacteria) begins to form at the centre. Pus is absent in wound healing unless secondary infection develops.

This process of inflammation is marked by *redness* and *warmth* due to increased vascularity, *swelling* due to exudation of fluid from the capillaries, and *pain* due to stretching of nerve endings.

As well as this local defence against invading organisms, there is a defence in depth against the foreign agents they produce, which operate throughout the body. This is the *immune response*, and is a reaction to foreign substances, usually proteins, known as *antigens*. The lymphocytes have complex recognition receptors on their surface membranes and the function of the immune system is to maintain chemical integrity, recognize cancer cells, damaged cells, microorganisms or alien molecules not genetically akin to the body, and to inactivate and eliminate them. There is a synthesis of gammaglobulins in the plasma, known as *antibodies*, which interact specifically with foreign antigens, neutralizing their toxic effects, destroying bacteria or facilitating their phagocytosis. This defence is necessary because bacteria may affect the body as a whole, either by invading the blood and lymph streams, or by forming *toxins* which circulate in the blood. Some toxins, such as those of diphtheria and tetanus, are extremely virulent and may be fatal, though the bacteria remain localized to the throat or wound.

In most cases the primary immune response is not immediate; but once antibodies are formed the indvidual is said to have an *acquired immunity*, since the antibodies either persist or can be renewed very quickly in subsequent attacks. Since this immunity is due to the individual's having overcome an attack of the disease by making the protective antibodies in his or her own tissues, this is called *acquired active immunity*. Active immunization may also be achieved by giving small doses of toxin or dead bacteria by inoculation or injection. The response is as great as after a genuine attack.

It is possible to produce actively acquired immunity to, say, tetanus, by injecting a horse with increasing doses of toxin until the animal is unaffected by what would have been a lethal dose. Its serum now contains a large amount of antitoxin. The injection of its serum into a human confers an *acquired passive immunity* – passive since the human's own tissues played no part in elaborating the protective agent. Such protection is only temporary and is only of value in treating or preventing early disease; it is of no value whatever for long-term protection.

Babies possess a congenital immunity aganst infections such as measles. This is passive, derived while in the womb from the blood of the mother, who has a permanent active immunity dating back to a childhood attack. Because the infant's resistance is borrowed, it lasts only a few months.

The production of antibodies which circulate in the blood is known as the *humoral* type of immune response. Another specific immune mechanism, the *cellular* response, is effected by small lymphocytes that recognize and attack certain antigens, and migrate towards them from the blood to protect the body. One or other response may predominate, but usually both are operant. The antibodies are produced by B lymphocytes (B for bone marrow) and phagocytosis is by the T lymphocytes associated with the thymus.

The reticuloendothelial system mediates both the early phagocytic response and the long-term immune reaction. It produces both antibodies and lymphocytes in the lymph nodes, tonsils and spleen, and in the reticulum of the bone marrow. In the latter the reticulum cells are also the precursors of red cells and polymorphs, i.e. they are *multipotent* and can develop into lymphocytes, polymorphs or red cells according to circumstances.

LYMPHATIC SYSTEM

This consists of lymphatic vessel and lymph nodes; the spleen is the largest mass of lymphoid tissue. The lymphatics form plexuses in all the tissues except the nervous system. They drain into trunks with one-way valves, allowing fluid to pass only towards the heart, accompanying major blood vessels. In the limbs the nodes are situated mainly at the bends of the elbow and knee and the groin and armpit. Others are sited close to the main viscera in the abdomen, in the mesentery, and along the aorta and inferior vena cava; others at the roots of the lungs and the bifurcation of the trachea deal wth inspired polluted air.

An individual node is pinkish-grey and bean-shaped (see Figure 16.3). Afferent lymphatics drain into them and efferent vessels leave, carrying lymph, lymphocytes and antibodies. New lymphocytes are constantly formed in the germinal centres and enter the blood or lymph draining the node; they also re-enter the node from the circulation.

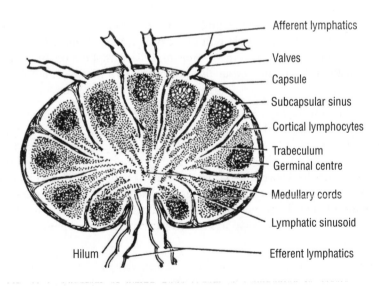

Figure 16.3 Diagram of lymph node (from *A Companion to Medical Studies*, Vol. 1)

The *lymph node* is a filter to which arriving lymph brings debris or bacteria for phagocytosis by reticuloendothelial cells. It also receives antigens and reacts by a combined humoral and cellular response. It produces new lymphocytes and is a staging centre for old ones.

The *spleen* (see Chapter 8) can be regarded as a very large specialized lymph node. Its reticuloendothelial content phagocytoses effete red cells and other debris. In certain conditions such as severe anaemia, red cells are, abnormally, produced there. It can contract and squeeze blood into the circulation to meet emergency demands, as in exertion, and it is also a source of lymphocytes and antibodies.

As it removes worn-out red and white cells and platelets, the spleen contains a good deal of iron pigment. Whenever red cells are increasingly destroyed, the spleen enlarges to cope, e.g. in malaria, where the parasite inhabits the red cells. Splenic enlargement is usual in countries where malaria is endemic.

However, the spleen is not essential to life. It is often damaged in accidents and can be removed without harm because its functions are taken over by other parts of the reticuloendothelial system. The lymphocytes make few or no antibodies, but in response to antigenic stimuli they become able to 'recognize' antigens and to crowd round them at local sites. Also, the lymphocyte can transform itself into a larger, more primitive cell that does form antibodies. Lymphocytic immunity is important in the reaction to grafts from other individuals. Tissue material from anyone except an identical twin is recognized as foreign and cast off. A badly burned child may be skin-grafted from its parent with apparent success, but the grafts slough off after a few weeks due to local lymphatic attack. This is because an individual contains antigenic substances common to no other person. Similar considerations apply to the transplantation of whole organs such as the heart or kidney, where the immune response may need to be suppressed by drugs for many months or years. All this relates to the *host versus graft* reaction. There is also a *graft versus host* reaction. Certain cell transfusions may damage the liver, lungs, kidneys and lymphoid system, most often transfusions of blood or blood fractions.

THYMUS

The thymus is a large lobed structure behind the upper part of the sternum, covering the great vessels arising from the heart and upper part of the pericardium (Figure 16.4). It is largest in early life and shrinks in the adult to a mere remnant. It is a lymphoid organ but does not share in the general lymphatic circulation. Many lymphocytes are formed here, but most never leave the gland.

In the foetus and infant the thymus seems to control the development of lymph nodes and lymphoid tissue. If it is absent these fail to develop, lymphocytes are few, and death results from infection because of failure of immune processes. In adults, the thymus is a source of fresh lymphocytes for immunologic imprinting, but it is not the only one and its removal causes little harm. After heavy irradiation the activities of bone marrow and lymphoid tissue are severely depressed, sometimes temporarily abolished. The thymus is more resistant and controls restoration of lymphocyte production and cellular immunity. This is important because irradiation is often used to suppress an unwanted immune response, e.g. after a transplant.

The thymus seems to be important in 'immunological surveillance', i.e. the capacity to recognize certain antigens and malignant processes as foreign and react against them. The converse of this is 'immunological tolerance'. An individual is tolerant of his or her own tissues and does not normally react against them. An individual is also tolerant of foreign antigens encountered in foetal life (and must be since he or she is within another person's body) and this tolerance extends into the neo-natal period. Sometimes, however, the immune response can be triggered by an individual's own tissue antigens and can attack them, resulting in

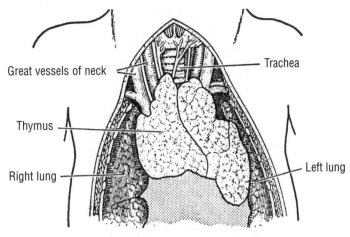

Figure 16.4 The thymus in the new-born child (after *Gray*)

auto-immune diseases such as thyroiditis or rheumatoid arthritis. This is due to uncontrolled activity of the T lymphocytes, which have evolved to recognize and destroy 'not-self' material but now turn to cannibalism. These disorders can be controlled by steroid drugs or other drugs which suppress immune reactions and must therefore be given after transplants. The thymus is actively involved in this process, in which a self-antigen comes to be regarded as hostile. If there is a long interval between the initial sensitizing dose and a later exposure, the protective reaction may be so violent as to endanger life. The neutralization occurs not in the bloodstream but in the tissues, and severe shock and collapse – even death – may occur. This reaction, known as *anaphylaxis*, happens rarely but occasionally in connection with such various agents as bee-stings or nuts.

IMMUNE-DEFICIENCY SYNDROMES

The defensive functions of the reticulendothelial system may be defective from birth as a *congenital immune-deficiency syndrome*. In these cases the child must be raised in a sterile enclosed environment to reduce the risk of infections. If the deficiency is *acquired,* the usual cause in the past was the action of immunosuppressant drugs such as steroids deliberately used to prevent the rejection of grafted organs, such suppression usually being temporary. AIDS is an acquired immunodeficiency syndrome brought about by HIV, the human immunodeficiency virus. This is transmitted by sexual intercourse, by inoculation, via transfusion of blood or blood products, or by transfer from mother to child in the womb. AIDS has a long incubation period, during which the host is potentially infective to others. Viral DNA is permanently incorporated into the DNA of host cells and when these foreign genes begin to dictate the formation of new viral particles, the host cells are destroyed. The virus is attracted to and destroys circulating lymphocytes of the immune system, but it also infects other types of reticuloendothelial cells in many organs and may directly attack the cells of the nervous system. Thus the clinical manifestations include:

▶ 'opportunistic' lung and other infections, as well as cancers due to failure of immunologic surveillance

▶ progressive dysfunction of the nervous system, lungs, bowel, skin and other organs.

Within five years of infection, about a quarter of untreated victims develop full-blown clinical disease. However, survival and an approximation to normal life has been provided by new drugs.

Test yourself

1 Which protein in blood is required for blood clotting?
   a Globulin
   b Albumin
   c Fibrinogen
   d Plasma

2 Where are red blood cells formed in children?
   a Liver
   b Spleen
   c Lymph nodes
   d Bone marrow

3 What is the main vitamin required for red blood cell production?
   a Vitamin A
   b Vitamin C
   c Vitamin D
   d Vitamin B12

4 Which blood group can give blood to any type of blood group?
   a O
   b A
   c B
   d AB

5 Which white blood cells use phagocytosis to remove foreign substances?
   a Polymorphs
   b Lymphocytes
   c Monocytes
   d Erythrocytes

6 Which proteins react to foreign substances?
   a Antigens
   b Antibodies
   c Amino acids
   d Antiglobulins

7 Injecting a person with antitoxin provides which of the following?
   a Acquired passive immunity
   b Acquired active immunity
   c Humoral immunity
   d Cellular immunity

8 Which of the following does not produce antibodies and/or lymphocytes?
   a Lymph nodes
   b Tonsils
   c Spleen
   d Liver

9 Which disease causes the spleen to produce red blood cells?
  a Rickets
  b Malaria
  c Chicken pox
  d Tetanus

10 Identify the location of the thymus.
  a Behind the sternum
  b In front of the lymph nodes
  c Behind the spleen
  d In the neck

17

The heart and circulation

In this chapter you will learn about:

▶ *the systemic and pulmonary circulations*

▶ *the heart*

▶ *nervous control of the heart*

▶ *the circulation*

▶ *the effects of exercise on the cardiovascular system.*

The circulation is a closed circuit, round which the blood is propelled by the contraction of the heart. Blood is driven into the *arteries*, which are thick-walled elastic tubes and by their recoil aid its distribution. The arteries divide into smaller *arterioles* and finally into a meshwork of *capillaries* – microscopic thin-walled vessels that pervade every tissue except the cornea of the eye, the outer layer of the skin and articular cartilage. This meshwork joins up again to form small *veins* which become larger venous trunks as they travel centrally towards the heart. The veins are thin-walled, have no pulse and contain valves to prevent any backward flow.

The arterioles communicate freely by cross-branches called *anastomoses* which ensure an adequate blood supply to any part; even when one vessel is obstructed there is a way round. In a few situations there are no anastomoses and the local artery is an end-artery, the only blood supply; the central retinal artery is an end-artery and its occlusion causes blindness.

The circulation is a device for transporting materials to maintain the internal equilibrium of the body tissues and fluids. Its functions are:

▶ to transport oxygen from lungs to body cells and carbon dioxide from cells to lungs

▶ to transport nutrients to the cells and wastes from cells to kidneys

▶ to transport excess heat to the skin to be lost to the exterior, so as to keep the body temperature constant.

The blood is also a vehicle for a number of hormonal control systems, e.g. the feedback mechanism between blood glucose level and insulin production, or between state of dilution and pituitary activity (Chapter 21).

There are two separate circulations:

▶ a *systemic* circulation concerned with the body as a whole and driven by the left side of the heart

▶ a *pulmonary* circulation concerned with passage of blood through the lungs and driven by the right side of the heart.

The two sides of the heart are separate, each with an upper chamber or *atrium* receiving blood from the great veins and a lower chamber or *ventricle* discharging blood into the great arteries. De-oxygenated venous blood from the body enters the right atrium, passes to the right ventricle, and is expelled through the pulmonary artery to traverse the lung capillaries. Here it receives fresh oxygen from the air sacs and gives up carbon dioxide to be exhaled. The fresh blood returns from the lungs in the pulmonary veins to the left atrium, then down to the left ventricle. It then leaves the heart via the great artery, the *aorta*, to the head, trunk and limbs. In the tissues the blood becomes dark and venous and is collected up into the great veins. The superior vena cava drains the head and arms, and the inferior vena cava drains the trunk and legs.

The arteries of the body contain bright blood and the veins dark blood; the reverse is the case for the pulmonary vessels because the lungs are concerned with reversing these chemical states.

There is a special arrangement of the abdominal vessels. Whereas the veins leaving most structures pass directly to the heart, those from the stomach and bowel enter the liver via the *portal vein*, which breaks up into a second set of capillaries so that the blood is filtered through the liver before reachng the heart. This is to ensure that the liver utilizes and stores foodstuffs digested and absorbed from the bowel and is known as the *portal system*.

Physiological features of the systemic circulation

The circulation is a closed system and, although the calibre of the blood vessels varies enormously, the blood never escapes into free contact with the tissues. The heart acts as a boosting mechanism in a pipeline, receiving blood from the veins at low pressure and pumping it to the arteries at a high one. When the blood eventually reaches the thin-walled capillaries, the plasma diffuses through their walls to mix with the tissue fluids which bathe the cells.

There is a *pressure gradient* from the high level in the great arteries to an intermediate one in the capillaries and still lower in the veins, which may be zero. Circulation in the veins largely depends on outside assistance such as the sucking action of respiration, which draws blood into the chest, and the contraction of the muscles around the veins in the limbs. Thus, on walking, the leg veins are emptied by the compressive action of the muscles, their valves preventing reflux and directing the flow to the heart.

This difference between veins and arteries is reflected in structure: the veins are thin-walled with little muscle or elastic tissue; the arteries have thick coats with much muscular and elastic tissue. The arterial tree is an elastic distributing system with a recoil to the heartbeat, which is transmitted to produce the pulse wave felt at the wrist and temple. The contractile power of the arteries maintains a higher blood pressure at a greater distance from the heart than would a system of inelastic tubes. The smaller arteries and arterioles can also exert a selective action on the circulation in particular regions by constricting or dilating to shut off or increase the local blood flow. The veins form an inelastic reservoir which may become a stagnant pool if their valves become inefficient or the heart fails.

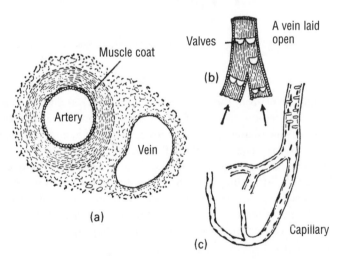

Figure 17.1 (a) Cross-section of an artery and vein, showing the much thicker muscular coat of the former; (b) vein slit open longitudinally, showing the valves with their pockets directed towards the heart; (c) a capillary with its delicate walls

The flow in the veins is slower than in the arteries because of their greater total cross-sectional area, though the volume of blood transmitted per minute must be the same. In the arteries there is friction between the viscous blood and the vessel wall and the blood cells tend to cling to the lining, so there is a more rapid central axial flow. This is one factor in the *peripheral resistance* to blood flow, which is also linked to the length of the vessels and their calibre.

$$\text{Blood pressure} = \text{cardiac output} \times \text{peripheral resistance}$$

As the arteries divide, the cross-sectional area of their branches increases until the total sectional area or *capillary bed* may be 1000 times that of the aorta. This is another aspect of the progressive fall in blood pressure as the arterioles branch into capillaries. When all the capillaries are widely dilated, as in shock, the circulation bleeds into itself and the blood pressure falls very seriously.

Venule

Arteriole

Figure 17.2 Diagram of a capillary network and its supplying and draining vessels (from *A Companion to Medical Studies*, Vol. 1)

The carotid sinuses and the walls of the main arteries in the neck and thorax contain *baroreceptors* sensitive to pressure changes. These act reflexly on the vasomotor centre in the hindbrain to produce vasodilatation if the pressure is too high, and vasoconstriction to raise a pressure that is too low. This is additional, or complementary, to the hormonal control exerted by adrenaline and other agents. The pressure is also regulated by kidney function, in the short term both by its effect on blood volume and by the internal secretion of *renin*, which stimulates aldosterone production by the adrenal cortex via a vasoconstrictor called *angiotensin*. Chronic kidney disease may cause permanent *secondary hypertension*, but the situation is complicated because primary, inherited or *essential hypertension* can itself also damage the kidneys in a vicious cycle.

The heart

The heart is a pump which continues to beat regularly and continuously for about 70 years with a rest interval never longer than a fraction of a second. Its rhythmic contraction is inherent and independent of nervous control. The transplanted heart has no nerve supply, and chicken heart cells have been kept contracting in culture media for many years.

The heart beats about 70 times a minute at rest, more rapidly in children, and up to a 150 times a minute in the child in the womb. Contraction is known as *systole* and relaxation as *diastole*. Each beat begins as a simultaneous contraction of both atria expelling blood into the ventricles. This is followed by ventricular contraction which throws the blood into the great arteries. This cycle lasts eight-tenths of a second, of which atrial contraction makes up one-tenth, ventricular contraction three-tenths, and diastole four-tenths. During diastole the atria act as passive receptacles draining the great veins. The atria contract a little before the ventricles, and do not have as much work to do as the ventricles so they are relatively thin-walled. The ventricles have the resistance of the peripheral circulation to overcome, notably the elasticity of the arteries, and are thick-walled and powerful. The right ventricle is only concerned with driving the blood through the lungs and therefore has less work to do than the left, so its wall is only a quarter as thick.

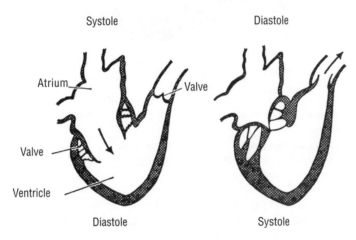

Figure 17.3 In relaxation, or diastole, the blood flows into the ventricle from the atrium through the open valve between

In contraction, or systole, the valve is slammed shut as the ventricle expels its blood into the aorta or pulmonary artery

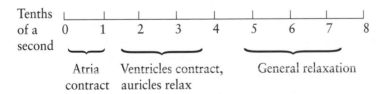

The regular sequence of the cardiac cycle is due to little foci of sensitive connective tissue which trigger each contraction. These *nodes* are stationed one near the junction of great veins and atria, one at the atrioventricular junction. They are connected to the heart muscle by a cable of fibres which run down the septum between the two halves of the organ (the bundle of His), giving a branch to either side (Purkinje fibres). The heart is flaccid in diastole; when it contracts it becomes thick and cone-shaped and transmits an impulse to the front of the chest on the left side.

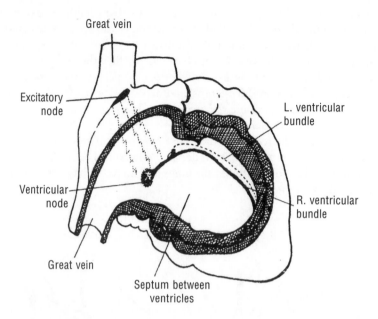

Great vein

Excitatory
node

L. ventricular
bundle

Ventricular
node

R. ventricular
bundle

Great vein

Septum between
ventricles

Figure 17.4 The right side of the heart, opened to show the mechanism which excites regular contraction of the muscular chambers

The stimulus begins at the excitatory node and spreads to the ventricular node, whence it is conveyed by a special bundle of fibres to each ventricle (adapted from Starling's *Principles of Human Physiology*)

This is the *apex beat*, between the fifth and sixth ribs. If one listens with a stethoscope or the unaided ear over this point, two characteristic sounds can be heard with each beat, coming closely together and repeated after the diastolic pase. These may be represented as follows:

...lubb, dupp...lubb, dupp...lubb, dupp...

and are both associated with ventricular contraction. The first is the slamming shut of the parachute-like valves preventing reflux from ventricles into atria. The second occurs at the end of ventricular contraction, when the semilunar valves at the mouths of the great arteries fall together to prevent backflow into the ventricles.

Contraction of the heart chambers must be coordinated. An uncoordinated contraction is known as *fibrillation*, which tends to prevent any proper pumping action. Atrial fibrillation is common, not serious, and controllable with drugs or electric shock. Ventricular fibrillation is rapidly fatal if not treated promptly.

Although the two halves of the heart constitute a single organ, they perform separate functions. The atrium and ventricle of each side constitute a pair of chambers concerned with pumping blood through a particular territory. Although the blood traverses each pump in turn, it is possible to imagine the right and left halves as separate organs fixed together for convenience (Figure 17.5). The output of the heart must obviously be the same on both sides or blood would be dammed up on one or other side. This output, at rest in the adult, is some 3–4 litres a minute, i.e. the whole of the blood in the body passes through the heart in under two minutes.

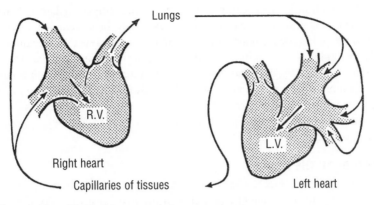

Figure 17.5 The two sides of the heart represented as separate organs

The *stroke volume* is the amount of blood ejected in systole from the left ventricle into the aorta, some 60–70 ml per beat. The *output* per minute is equal to this quantity multiplied by the pulse rate.

CONTROL OF THE HEART

The heart's activity is varied by altering either the *heart rate* or its power of contraction, i.e. its *output*. The factor most influencing output is the inflow from the great veins, for output must equal intake whatever the resistance of the systemic circulation, or blood will be dammed up in the venous system. Venous inflow is greatly influenced by respiration. Breathing in produces a negative pressure in the chest which draws blood into the great veins and right atrium from the venous trunks of the head, abdomen and limbs. Therefore, filling of the heart is largely due to ordinary breathing, a deep breath increases intake, the output rises correspondingly, and the pulse quickens. Conversely, if pressure within the chest is raised by straining, the great veins are flattened, cardiac output falls, and the pulse weakens.

Mechanically, the work of the heart is used in two ways:

1 blood expelled acquires a definite velocity and momentum

2 the resistance of the arterial tree has to be overcome, the stretching of the arterial walls and their tendency to recoil providing a store of potential energy like a compressed spring.

During the course of a normal life the work done by the heart is equivalent to lifting a weight of 10 tonnes through a height of 16 km.

The circulatory resistance is equal to the arterial blood pressure. The heart adapts to expel exactly its intake of blood whatever the arterial pressure, or to expel a varying intake against a constant resistance. This adaptability lies in the power of heart muscle fibres to contract more forcibly the more they are stretched, and vice versa. Obviously, there are limits to the ability to meet increased demand by dilating, with increased contraction and hypertrophy of its fibres. One limit is that its fibrous envelope, the pericardium, cannot stretch. The other is the mechanical disadvantage at which stretched fibres have to work. This property of overcoming handicaps such as a leaking valve or raised blood pressure is known as *compensation*, and results in an organ larger and thicker than normal. But when the limits of adjustment are surpassed, *decompensation* results; output cannot keep pace with influx, the veins become stagnant reservoirs, and the soft tissues become waterlogged.

The blood pressure within the heart can be measured by inserting a cardiac catheter, a tube passed from an artery or vein into the organ. The pressure in the systemic circulation is five times that in the pulmonary circuit. Stabilization of the blood pressure is important, particularly for the kidneys and brain. The kidneys stop secreting urine if the systolic pressure falls below 50 mm of mercury; and one cannot survive unharmed if the brain is deprived of arterial blood for more than a few minutes. Even if this is not fatal, deprivation of flow to the cerebral cortex while that to the vital centres in the medulla persists causes *brain death*, a permanent vegetative state. The organism as a whole continues to function but consciousness and volition are permanently lost.

NERVOUS CONTROL OF THE HEART

Apart from its self-regulating mechanism, there is a nervous control to adapt the organ to the needs of the body. This is effected by branches of the two constituents of the independent autonomic nervous system – the sympathetic and parasympathetic. As with the bowels and bladder, the two are mutually antagonistic but work together like a pair of reins to achieve a balanced control (Figure 17.6).

The *parasympathetic* fibres are derived from a nerve called the *vagus*, running from the base of the brain through the neck and chest. They depress heart action, slowing its rate and weakening its force. If the nerve is stimulated the heart may stop; firm pressure on the eyeball does this and slows the pulse. Sudden death may occur from *vagal inhibition* as when diving into cold water. The vagus acts by liberating *acetylcholine* at its branch endings in the heart muscle.

The *sympathetic* fibres are derived from the sympathetic chain which runs on each side of the spinal column. They accelerate the heartbeat and increase contraction force by liberating adrenaline. Adrenaline also reaches the heart through the bloodstream during stress.

The ultimate cell stations from which both types of fibres derive are in the hindbrain, or medulla (Chapter 20), where they form two centres. One inhibits and the other augments cardiac function. These exert central control on the basis of information received from various parts of the body, particularly from stretch or pressure receptors in the heart and great arteries. These messages indicate the filling of the heart, blood pressure and extent of cardiac dilatation. The control centres react to secure that the heart beats faster and more forcefully if venous influx increases or the arterial pressure falls, and slower and more weakly if the changes are reversed.

The heart gets its own blood supply from two tiny vessels, the *coronary* arteries, which spring from the aorta at its origin from the left ventricle and ramify over the surface of the organ. These are very sensitive to nervous control and changes in oxygenation of the blood, and increase the blood supply to the heart wall when extra demands arise. Coronary disease (atherosclerosis) or obstruction by clot (thrombosis) interfere with the heart's capacity to respond to demand and may cause sudden death of part of the muscle wall (infarction). It is now commonplace to replace diseased coronaries by vein grafts or to expand constricted segments by balloon catheterization.

The great energy output of the heart is obtained from free fatty acids, ketoacids and lactate in the blood. The heart muscle is sensitive to changes in concentration of sodium, potassium and calcium ions.

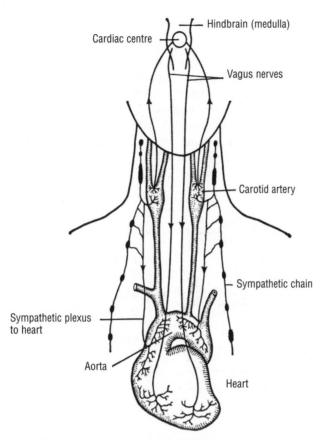

Figure 17.6 The nervous control of the heart; anatomical arrangements (adapted from Starling's *Principles of Human Physiology*)

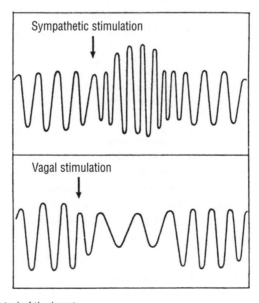

Figure 17.7 The nervous control of the heart
Sympathetic stimulation produces a rapid forceful beat, parasympathetic action a slower beat of less amplitude

The circulation

There is a *pressure gradient* in the circulation, falling from the left ventricle through the arterioles and capillaries to very low or zero in the veins. This pressure is measured in terms of a column of mercury. In the great arteries it is about 130 mm but falls to 20 mm in the capillaries and only a few millilitres in the great veins. In the large veins of the neck and chest the pressure is often negative because of the sucking action of respiration. Wounds of the veins of the neck may suck in air from outside, which is dangerous as it gets churned in the heart and obstructs blood flow. When cut, arteries bleed from the end nearer the heart in spurts. Veins bleed from the distal end in a gentle flow.

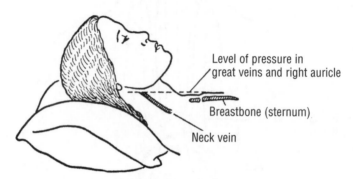

Figure 17.8 The basic level of venous pressure in the reservoir of the right atrium of the heart is shown by the extent of the visible column of blood in superficial veins of the neck

The blood pressure in the arteries is alternately high and low, correspnding to contraction and relaxation of the left ventricle. The *systolic pressure* is some 120 mm of mercury, the *diastolic pressure* only 70 mm; the fluctuation between is the *pulse pressure*, i.e. 60 mm. In the veins and capillaries the pressure is steady. In common usage the term 'blood pressure' refers to arterial pressure and is expressed in double figures as 120/70, though the range is great. Since the higher systolic pressure depends on contraction of the left ventricle, and this is influenced by emotion and exertion, it is the steady diastolic pressure that provides the base line.

The intermittent flow from the heart has become a constant flow when the capillaries are reached, due to the recoil of the arteries and arterioles which continues the driving force while the heart itself relaxes.

The time taken for the blood to pass from any given point through the heart and lungs and back again is known as the *circulation time,* and is about 25 seconds at rest. This does not imply that the speed of the blood is constant, as this varies with heartbeat and the size of the blood vessel. At its maximum, in the arteries, the blood velocity is about a mile an hour. We must not confuse the velocity of blood flow with that of the *pulse wave*. The pulse wave is a pressure front transmitted at some 24 kph (15 mph) through the arterial blood and is unrelated to the forward movement of the fluid itself.

The network of tiny arterioles and capillaries, or capillary bed, provides the main resistance to left ventricular thrust. But it is not a fixed resistance, for the arterioles have muscular coats and the capillaries contractile cells, which are subject to nervous and chemical control. *Vasoconstriction* is when the blood vessels are narrowed, and *vasodilatation* is when they open up. This allows adjustment of the dynamics of the circulation in different parts, an adjustment needed to meet the changing effects of gravity in different postures.

In standing, the whole of the blood would collect in the small vessels of the abdomen and legs were these not largely shut down by vasoconstriction. This causes a compensatory rise of blood pressure throughout the body. Conversely, extensive vasodilatation leads to stagnation and a fall in blood pressure. The enormous vascular area of the abdominal organs forms the *splanchnic pool*, in contrast to the *somatic* capillary bed in the skin, muscles and skeleton. When the splanchnic vessels are dilated, as after a heavy meal, the blood pressure falls and the supply to the brain is reduced. We become sleepy and tend to lie down because of a temporary cerebral anaemia and the circulation is maintained more easily in the horizontal position. Fainting is an acute change of the same nature after a shock that produces splanchnic vasodilatation.

Often, in older individuals, this control mechanism fails and there is a sudden fall in blood pressure, or postural hypotension, on standing, which may cause falls or fainting attacks. In *surgical shock*, which develops after injuries or burns, there may be such widespread expansion of the capillaries that the body 'bleeds to death into its own blood vessels'. The pressure falls profoundly, the blood is more viscous, and too little blood is available for the heart to work on or the lungs to aerate.

The infusion of fluid into the blood has little lasting effect if it is crystalloid as the excess water and salts pass rapidly into the tissue spaces. If the fluid has a colloid content, this cannot happen and the vessels accommodate the added fluid, pressure is restored, and the kidneys are stimulated to pass more urine. Loss of blood, as in donation of, say, 600 ml, is compensated so efficiently that the circulatory effects are negligible. But a sudden loss of several pints causes a severe fall of blood pressure, which must be compensated. There is general vasoconstriction, the heart beats faster, sweating occurs, the skin becomes cold, kidney blood flow is reduced and urinary output falls. At the same time fluid enters the circulation from the tissue spaces and the blood pressure and volume are restored, though the blood is more dilute and full cell count may take some days or weeks to be restored.

The circulation is like a railway system with numerous sidings but a constant number of trains. If some of these are shunted off, the main line traffic will be less heavy, i.e. the blood pressure falls. If all the trains are in use, the traffic is intense and the pressure rises. Any inflammation sets up a local vasodilatation which flushes that part of the body with blood; the arterioles may be so widely opened that they communicate their pulse to normally pulseless capillaries resulting in that body part throbbing.

The mechanism controlling the small vessels is like that of the heart: part nervous, part chemical. The sympathetic fibres that accelerate the heart constrict the vessels and prepare the body for emergencies by diverting blood from the skin and non-essential organs to the muscles and brain. The parasympathetic promotes relaxation and fall of blood pressure. But the two systems are reciprocal; the vessels are never completely shut down or completely dilated, except in severe shock.

The arterioles and capillaries are very sensitive to circulating adrenaline, which has the same action as the sympathetic system, and are dilated by substances liberated from damaged tissues, e.g. *histamine*. The ultimate control centres are in the medulla, and their prime function is to safeguard the blood supply to the brain. Other tissues can survive deprivation of oxygen for an hour or more, but the brain cells die after more than a few minutes of lack of blood flow. This is seen after head injury causing progressive bleeding within the skull, which compresses the brain and makes it more difficult for blood to be pumped in from the heart. The vasoconstrictor centre in the medulla reacts by shutting down vessels all over the body so that blood pressure may rise to enormous heights to force blood into the brain-box, at least until the process collapses.

EXERCISE (SEE ALSO CHAPTER 23)

Although the heart and vessels adjust to demands, they are anticipated by mental factors. The emotion in waiting to start a race causes a rise in blood pressure and pulse rate long before these are physically necessary.

During strenuous exercise, the oxygen demands of the muscles may be ten times those of the resting state, and this must be met by increased cardiac output. Arterial blood is already saturated with oxygen and cannot dissolve more. The only way to bring more oxygen to the tissues is to rush the blood carrying it round the circulation at a faster rate.

Bodily changes in exertion are complex. Muscular contraction speeds return of blood to the heart by milking it upwards in the veins. This increased influx imposes an equivalent increase on cardiac output if the heart is not to fail. The amount of blood in the circulation cannot be increased, but more is diverted to the muscles by shutting down the vessels of the skin and splanchnic area and by contraction of the spleen. Exaggerated respiratory movements suck more blood into the great veins. The sympathetic system, aided by outpouring of adrenaline into the blood, increases heart rate and the force of each beat, and mobilizes liver glycogen as sugar for the muscles to burn. The waste products of muscle contraction – carbon dioxide and lactic acid – stimulate the brain centres controlling the heart, vessels, and the depth and rapidity of respiration.

We note elswhere that short periods of light exercise are associated with *aerobic* respiration, which consumes oxygen and burns glucose to carbon dioxide, water and most of the ATP (adenosine triphosphate) required. In continued heavy exertion the circulation cannot deliver nutrients to the muscles fast enough and *anaerobic* respiration takes over. Glycogen is converted to ATP and lactic acid without using oxygen, leaving an *oxygen debt* which must be paid off by deep breathing for a period after exertion to restore biochemical normality.

Test yourself

1 Which statement about the structure of arteries is correct?
- **a** They contain valves
- **b** They have thin walls
- **c** They have thick walls
- **d** They carry blood under low pressure

2 Which receptors in the neck are sensitive to blood pressure changes?
- **a** Mechanoreceptors
- **b** Chemoreceptors
- **c** Baroreceptors
- **d** Proprioreceptors

3 Which is the main nerve in the parasympathetic nervous system?
- **a** Vagus
- **b** Aorta
- **c** Vena cava
- **d** Carotid

4 Which blood vessels provide the heart with its own blood supply?
- **a** Aorta
- **b** Pulmonary arteries
- **c** Coronary arteries
- **d** Hepatic vein

5 What is the stroke volume at rest?
- **a** 30 ml
- **b** 40 ml
- **c** 50 ml
- **d** 60 ml

6 What is the average adult's systolic blood pressure?
- **a** 120 mm of mercury
- **b** 90 mm of mercury
- **c** 70 mm of mercury
- **d** 50 mm of mercury

7 Which of the following is a hormone of the sympathetic nervous system?
- **a** Insulin
- **b** Cortisone
- **c** Adrenaline
- **d** Dopamine

8 What name is given to a blood clot?
- **a** Thrombosis
- **b** Infarction
- **c** Atherosclerosis
- **d** Vagal inhibition

9 What makes the characteristic 'lubb, dupp' sounds of the heart?
   a   Blood
   b   Heart valves closing
   c   Fibrillation
   d   Diastole

10 Which of the following supplies the heart with energy?
   a   Carbohydrates
   b   Fatty acids
   c   Protein
   d   Cholesterol

18

Respiration

In this chapter you will learn about:

▶ *the respiratory system*

▶ *respiration*

▶ *lung ventilation and lung volumes*

▶ *oxygen, carbon dioxide and their effects on respiration*

▶ *tissue respiration*

▶ *the regulation of respiration*

▶ *the effects of oxygen deficiency, altitude and depth on breathing.*

The respiratory system

Inhaled air enters the larynx and travels through the air passages proper. The first part is the *larynx*, which leads to the *trachea* (windpipe) which divides in the upper chest into a right and left *bronchus* for each lung. Each bronchus subdivides in the lung into numerous *bronchioles* ending in clusters of tiny *air sacs* or *alveoli*. It is in their walls that interchange occurs between the gases dissolved in the blood and those of the inhaled air (Figure 5.10). The alveolar walls are coated with a liquid film of *surfactant* which lowers surface tension and facilitates lung expansion.

The chest cavity is divided into right and left by a partition, or *mediastinum*, in which the heart is embedded. The two halves are separate and contain the lungs. Each cavity is lined by the smooth membrane of the *pleura*, and each pleural space is a closed sac as the membrane is reflected from the chest wall to cover the surface of the lung. Normally, there is no pleural cavity as the two layers are in contact, and each lung fills its side of the chest. But the lung is very elastic, tending to shrink and expel its contained air; since in health it cannot do this, there is a negative pressure in the potential pleural space. The moment air enters, the lung collapses into a small, solid, airless mass.

Respiration

The purpose of respiration is the entry and exit of air to and from the lungs to supply the body with oxygen (O_2) to burn foodstuffs in the tissues and to eliminate waste carbon dioxide. In *inspiration* the chest cavity is enlarged and air enters; in *expiration* the reverse occurs. Respiratory movements may be thoracic or abdominal, or both. In *thoracic inspiration* the sternum is lifted, the ribs are elevated by the intercostal muscles and lie more horizontally, and the chest increases in diameter from side to side and front to back. In *abdominal inspiration* the diaphragm contracts down on the abdominal organs, bulging the abdominal wall and increasing the vertical height of the chest. Inspiration is active, due to muscular contraction. Expiration is passive; the chest wall subsides, the abdominal muscles recoil, the diaphragm relaxes and rises, and air is driven out of the lungs.

LUNG VENTILATION

During quiet breathing about 600 cm³ of air are taken in and breathed out at each respiratory cycle; this is the *tidal volume*. On deep inspiration, some 2000 cm³ more can be taken in, the *inspiratory reserve volume*; and after a normal expiration it is possible forcibly to expel a further 1200 cm³, the *forced expiratory volume*. The sum of these figures is the total possible flow in one direction, the *vital capacity*, and represents the maximum volume of air that can be expelled after as deep an inspiration as possible:

Inspiratory reserve volume	=	2000 cm³
Tidal volume	=	600 cm³
Expiratory reserve volume	=	1200 cm³
Vital capacity	=	3800 cm³

The vital capacity varies a good deal, depending on physique and fitness. It is greatest when standing and least when lying down, and is diminished when the ribs are fixed by arthritic disease. But no expiration, however forcible, can empty the lungs of air. A minimum volume of air is left in the bronchi and air sacs, the *residual volume*, some 1500 cm³. Finally, not all the 600 cm³ of tidal air reaches the air sacs since some is needed to fill the air passages themselves, a *dead space* of some 150 cm³.

The air passages are not rigid tubes. They have a strong muscular wall which can alter their calibre by constriction or relaxation, so varying the volume of the dead space and the resistance to airflow. *Asthma* is an intense spasm of bronchial muscle and makes expiration a long drawn-out effort; however inspiration is less affected. The smooth muscle of the bronchi is controlled by the parasympathetic and sympathetic nerves. The sympathetic nerves, in exertion and fright, open the airways widely to achieve maximum possible influx. Consequently, adrenaline, which stimulates sympathetic nerve endings, is injected to combat an acute asthmatic attack.

Chemical aspects of respiration

In the tissues some 400 cm^3 of oxygen is used every minute to burn foodstuffs, with the production of carbon dioxide. The oxygen is taken from the blood in the capillaries and the carbon dioxide returned in exchange. Venous blood leaving any part of the body contains less oxygen and more carbon dioxide than the arterial blood that entered it. The lungs reverse this state in the venous blood, restoring its oxygen quota and expelling excess carbon dioxide by interchange in the air sacs. The venous blood is rendered arterial in the lung capillaries and the changes in the tissues are reflected in changes in the composition of the tidal air, which, when breathed out, contains less oxygen and more carbon dioxide than on inspiration.

	Oxygen	Carbon dixoide	Nitrogen
Inspired air	20.95%	0.05%	79%
Expired air	16.5%	4.0%	79.5%

The nitrogen content appears slightly greater in the expired air because slightly less air is breathed out than in. This is because the volume of carbon dioxide exhaled is slightly less than that of the oxygen for which it is exchanged. This ratio, or *respiratory quotient*:

$$\frac{\text{carbon dioxide output}}{\text{oxygen consumption}}$$

is always less than unity and usually about 0.85 and depends on the nature of the foodstuffs being burnt (see Chapter 13).

The respiratory gases are carried dissolved in the blood plasma (carbon dioxide) or loosely combined in the red cells (oxygen) and the total gas that can be expelled is about 70%. Of this, the relatively insoluble nitrogen forms only a small proportion. Analysis of venous and arterial blood shows their different gas content in volumes per cent:

	Oxygen	Carbon dixoide	Nitrogen
Arterial blood	19%	50%	1%
Venous blood	10.5%	58%	1%

The problem is not a simple matter of relative solubilities, for it is affected by the following factors.

1 The amount that can be dissolved is proportional to the pressure of the particular gas in the lungs or tissues.

2 The gases are not held in simple solution but in loose chemical combination: oxygen with haemoglobin in the red cells, carbon dioxide in the carbonic acid and sodium bicarbonate of the plasma.

3 The two gases tend to displace each other from their combinations with corpuscles or plasma.

OXYGEN

When venous blood, with its reduced oxygen content at low pressure, reaches the lungs, it is exposed to higher oxygen pressure in the alveolar air. The pressure gradient is towards the blood and oxygen combines loosely with haemoglobin to form oxyhaemoglobin. When the blood has been swept through the heart to the tissue capillaries, it arrives at a region where oxygen has been used in combustion and is at a lower pressure than in the blood. This gradient sets the other way and oxygen is set free and diffuses into the tissues. This dissociation is catalysed by the presence in the tissues of the excess carbon dioxide formed by combustion. The very waste product formed by using up oxygen stimulates the liberation of oxygen from its bound state in the blood.

CARBON DIOXIDE

The pressure of this gas as a waste product in the tissues is much higher than in the fresh blood from the arteries. The gradient favours passage of the gas from the tissue fluid into the capillaries, where it enters into loose combination as carbonic acid and sodium bicarbonate in the plasma. This is carried with the venous blood through the right heart to the lung capillaries, where it is exposed to freshly inhaled air with a very low carbon dioxide content. This gradient sets the other way and the gas is expelled with the exhaled air. This transference is assisted by the saturation with oxygen in the lungs, as a high oxygen content in blood tends to break up loose carbonic compounds.

Tissue respiration

We have been describing 'external respiration', concerned only with gas interchange in the lungs and tissue capillaries, and not with the vital processes in the tissue cells. This less obvious process, called 'internal respiration', is the motive of all the complex apparatus of heart, lungs and circulating blood. Every tissue uses oxygen and liberates carbon dioxide for its energy needs in amounts depending on functional activity, greatest in the tissues of the heart, brain, liver and muscles. Respiration involving the use of oxygen is known as *aerobic*. Many bacteria and other organisms respire without using oxygen, though still forming carbon dioxide, an *anaerobic* process. The fundamental processes in the cell between the intake of oxygen and the formation of carbon dioxide are governed by a complex system of enzymes known as *oxidases* that facilitate oxygen transfer and *dehydrogenases* that achieve an essentially similar result by removing hydrogen.

The newly arrived nutrients are elaborated by anabolism into molecules of cell constituents: lipids, proteins and glycogen. Food fuels, especially glucose, are broken down and the freed energy captures ATP, which is the driving force directing this energy to yield work. Nutrients and cell constituents are simultaneously broken down by the essentially catabolic activity of the mitochondria. These organelles use oxygen in completing breakdown of nutrients to carbon dioxide and water by way of pyruvic acid and acetyl compounds, and acquiring ATP. This last process is marked by *phosphorylation*, an oxidative process that exploits molecules for activity, motion or work.

Essentially, the objective of cellular respiration consists of glucose formation and the formation of ATP, an oxidative process necessarily coupled with the reduction of other substances, which fixes some of the chemical energy of the food molecule in a high-energy linkage. Food fuels such as glycogen store energy in the tissues and these stores are subsequently mobilized to produce ATP for cellular processes. The general equation for glucose breakdown is as follows:

$$C_6H_{12}O_6 + 6O_2 = 6H_2O + 6CO_2 + 36 \text{ ATP} + \text{heat}$$

$$\text{Glucose} + \text{oxygen} = \text{water} + \text{carbon dioxide} + \text{ATP} + \text{heat}$$

but the actual process, far too complex to be described in detail here, includes the Krebs cycle and electron transport chain.

REGULATION OF RESPIRATION

Respiration is complex, involving the nostrils, larynx, chest wall, diaphragm and abdomen, and accurate coordination is necessary. This is achieved by a *respiratory centre* in the medulla, one of a group of vital centres that includes the cardiac and vasomotor centres. It is constantly influenced by information from the lungs and elsewhere, and sensitive to changes in the oxygen and carbon dioxide content of the blood in the cerebral vessels, so as to alter respiration as needed. This is automatic and unconscious, though we can control breathing up to a point if we wish. It is one of the most fundamental of activities.

Nugget

In progressively deep anaesthesia or poisoning, respiration is one of the last bodily functions to cease.

The *chemical control* of respiration is effected by response of this centre to changes in the blood gases, i.e. their *partial pressures*. (The total pressure exerted by a mixture of gases is the sum of the independent pressures of each gas in the mixture, and the partial pressure of each is directly proportional to its percentage in the overall gas mixture.) It is most important to get rid of the carbon dioxide and other wastes. Whereas the oxygen content of the air is ample for respiration within wide limits, the slightest rise in carbon dioxide increases respiration. If carbon dioxide reaches an increase of only 1.5% it increases ventilation by half. In progressive asphyxia, with accumulation of the gas, respiratory convulsions develop, followed by collapse and death. There is much less sensitivity to oxygen lack, and the atmospheric level can fall from 21% to 13% without noticeable effect. Even at this point there is no great distress until collapse occurs without warning, as in unmasked pilots at high altitudes.

The *nervous* or *reflex control* is effected by the vagus nerves, which link the medullary centres with the heart and lungs. On the sensory side, these provide information on the distension of the lungs, blood pressure and gas content in the large arteries. On the motor side, they mediate reflex responses, modifying the action of the respiratory muscles.

Of the two parts of the respiratory centre, the inspiratory is dominant. Inspiration is active, due to messages sent down the vagus. Expiration, however, is a passive relaxation. The coordination of the two is an example of reflex action and feedback. When inspiration is at its height, the lungs are inflated and their sensory nerve endings stretched. These stimulate the expiratory part of the nerve centre, vagus action is inhibited, and expiration follows. The very act of inspiration produces the stimulus for the succeeding expiration.

The nervous control of respiration affects its rhythm, the chemicals control its depth, and these are reciprocal. If increased carbon dioxide increases the volume of inspiration, it must also accelerate the stretch response and thus the nervous control of rhythm. In exercise, the rising carbon dioxide in the blood as a waste reaches the brain and at first stimulates respiration. But as effort increases it is washed out of the blood by forced breathing, and the stimulus now comes from the lactic acid formed by muscle contraction and from secreted adrenaline. This steady state is the 'second wind'.

Nugget

Deliberate overbreathing at rest so empties the blood of carbon dioxide that the normal stimulus to breathing is removed. There is then no incentive to breath for the next couple of minutes, but the experiment is a little dangerous.

OXYGEN DEFICIENCY

A lack of available oxygen may be due to:

▶ lowered content in the inspired air

▶ obstruction of the airways

▶ lung disease

▶ inability of the blood to carry its full oxygen quota, as in haemorrhage or anaemia

▶ the haemoglobin being put out of action by a poison such as *carbon monoxide*

▶ the circulation being slowed in heart disease or inability of poisoned tissues to use oxygen.

All these result in overbreathing at rest, an effort to make up for the deficiency by increasing air intake. The cardiac invalid in his or her chair may pant as strenuously as someone who has run a mile. In each of these states the blueness of insufficiently oxidized haemoglobin is evident in the lips and under the nails – the condition of *cyanosis* (except in carbon monoxide poisoning, when the blood is cherry red).

HEIGHTS AND DEPTHS

Similar changes occur at altitudes; the available oxygen decreases as atmospheric pressure falls and reaches the danger point of 11% at 4877 m (15,000 feet) above sea level, above which an oxygen mask is desirable. Immediate compensation is by overbreathing and by an increased rapidity of the circulation. More permanent adaptation can be made by mountain dwellers, who acclimatize by an increase in the number of red corpuscles and habituation of the tissues to lowered oxygen content.

In working at depths or in compressed air, the pressure of carbon dioxide is increased proportionately and produces its usual toxic effects. At the same time nitrogen, normally poorly soluble in blood, is driven into solution and, if the pressure is suddenly released, is liberated in the blood as bubbles which may block vessels of the heart or nervous system and cause death or paralysis. Hence the necessity for recompression and gradual decompression.

Test yourself

1 Identify the action of the diaphragm in expiration.
   a   It contracts
   b   It relaxes
   c   It expands
   d   It lowers

2 What is tidal air?
   a   Air breathed in and out in each cycle
   b   Maximum air breathed in in one breath
   c   Maximum air breathed out in one breath
   d   Total volume of air breathed out in one breath

3 Which part of the respiratory system is affected by asthma?
   a   Alveoli
   b   Bronchioles
   c   Larynx
   d   Trachea

4 How is carbon dioxide transported in the blood?
   a   By white blood cells
   b   Attached to haemoglobin in red blood cells
   c   In platelets
   d   Dissolved in the blood plasma

5 Which product catalyses the dissociation of oxygen from haemoglobin?
   a   Oxygen
   b   Nitrogen
   c   Carbon dioxide
   d   Lactic acid

6 How can overbreathing stop a person from breathing?
   a   Excessive intake of carbon dioxide
   b   Excessive intake of oxygen
   c   Excessive intake of nitrogen
   d   Excessive intake of carbon monoxide

7 What condition occurs if a person has blue lips due to insufficiently oxidized haemoglobin?
   a   Anaemia
   b   Haemorrhage
   c   Cyanosis
   d   Hypothermia

8 Which organelles use oxygen to break down nutrients to produce ATP?
   a   Lymphocytes
   b   Ribosomes
   c   Endoplasmic reticulum
   d   Mitochondria

9   What is the function of a dehydrogenase enzyme?
    a   To remove hydrogen ions
    b   To facilitate oxygen transfer
    c   To catabolize glucose
    d   To break down fatty acids

10 What is a person's usual respiratory quotient?
    a   0.5
    b   0.65
    c   0.7
    d   0.85

The kidney and excretion

In this chapter you will learn about:

▶ *the kidney*

▶ *urine*

▶ *filtration*

▶ *tubular absorption and concentration*

▶ *the pituitary gland and diuresis*

▶ *kidney hormones*

▶ *the bladder.*

The kidney

The gross structure of the urinary tract is set out in Chapter 5. Microscopically, each kidney contains over a million secretory units, or *nephrons*. Each nephron comprises a tuft of blood capillaries, or *glomerulus*, invaginated into a capsule leading to a convoluted tubular system which discharges with others into a larger collecting tubule draining at the apex of a pyramid into the renal pelvis (Figures 8.19 and 8.20). Such is the functional reserve that two-thirds of one kidney will carry out the excretory function essential to life.

Each tubular system is closely enveloped by a plexus of blood vessels derived from a branch of the renal artery. From this the blood is gathered into a tributary of the renal vein, which removes the blood after purification. The kidneys are essentially a pair of filters through which approximately 1 litre of blood circulates each minute, i.e. the whole of the blood in the body passes through them in five or six minutes, or 1800 litres a day, which is 400 times the blood volume.

The general function of the kidney is to maintain a constant internal environment. It is a very efficient mechanism for preserving a constant chemical and physical composition of the blood plasma and extracellular fluid. The kidneys' role is to regulate:

▶ water content

▶ pH

▶ osmotic pressure

so as to maintain electrolyte and water equilibrium. The blood and tissue levels of mineral ions, organic metabolites and wastes must be kept within narrow limits to preserve health. Removal of the kidneys, or their failure, causes accumulation of urinary constituents in the blood (*uraemia*) with death after two to three weeks unless relieved by dialysis.

Practically all the waste products of metabolism are excreted in the urine. The only exceptions are:

▶ carbon dioxide, exhaled from the lungs

▶ the breakdown products of bile and other digestive juices, expelled in the faeces.

If the blood contains an abnormal non-colloidal constituent, or an excess of normal constituents such as water and salts, the kidneys excrete these to restore normality. They are the only means of eliminating the end-products of protein metabolism. But they do not themselves form any of the wastes they excrete; they merely rid the blood of substances formed in other parts of the body. In renal failure this process is inefficient, the blood levels of urea and other wastes rise enormously, and the tissues are poisoned by them. To fulfil these functions the kidneys have a huge blood supply.

Urine

The water content of the urine is derived partly from ingested food and fluids and partly from oxidation processes in the tissues. So its exact composition varies with the food, fluid intake, and fluid loss by other channels. Nevertheless, it remains fairly constant.

It is a clear yellow, paler when diluted by excessive drinking, darker when concentrated by sweating or dehydration. It contains mucus derived from the lining of the urinary channels and is moderately acid. The total output in 24 hours varies with fluid intake and outside temperature: it averages 1.5 litres with a range of 1–2.5 litres, and its formation is lowest in sleep and maximal during the day or on exertion. Its specific gravity varies around 1.01–1.03.

The urine may be turbid from calcium phosphate. It may deposit mucus, urates, crystals of calcium oxalate or phosphate. Microscopically, deposits include occasional casts shed from the lining of the renal tubules, sometimes red cells after exertion. Urine is frothy when shaken due to the presence of bile salts.

Its constituents are organic and inorganic. The *inorganic* group includes chlorides, phosphates, ammonia, sulphates, magnesium, calcium and iron. The ammonia is derived from protein breakdown, sulphates from oxidation of sulphur in protein, phosphates partly from the food and partly from oxidation of organic phosphates in the tissues. The *organic* components are mainly nitrogenous end-products of protein metabolism: urea, uric acid, creatinine and others. Urea is the most important and the urea content is a good index of protein metabolism. There is an excess of uric acid in gout, derived from nucleoproteins. There are also various organic pigments, derived from the bile. The non-nitrogenous organic constituents include oxalic acid from the food (e.g. rhubarb), lactic acid (after exertion) and derivatives of water-soluble hormones and vitamins.

Certain *abnormal* constituents appear at times. If the kidneys are diseased, some of the normal plasma proteins enter the urine, which coagulates on warming. The urine usually contains very little protein but it can occur in certain conditions, such as in pregnancy and after long standing in some people. Glucose is found in the urine of diabetics, lactose in nursing mothers, and ketone bodies in severe diabetes or starvation. There may also be haemoglobin, after an incompatible blood transfusion or in malignant malaria (blackwater fever), or increased bile salts and pigments in liver disease and jaundice.

FILTRATION

Urine formation entails the passage of water and crystalloids from the capillaries of the glomerulus into the nephron capsule and collecting tubules. Colloids, such as the blood proteins, do not normally appear in the urine. The essence of renal activity is that inorganic salts and organic compounds such as urea and uric acid are *concentrated* in the urine. The water content of urine and blood is similar, but there is 60 times as much urea in urine as in blood, 15 times as much uric acid, 40 times as much phosphate, seven times the amount of potassium, and twice as much chloride.

The basic processes are those of *filtration* in the glomeruli and *reabsorption of water* into the tubules. The energy required for filtration is considerable, as the osmotic pressure of the plasma proteins tends to hold the excreted substances in the circulation. The driving force that overcomes this is the blood pressure in the glomerular arterioles. Should this fall below the osmotic pressure the blood proteins (40–50 mm of mercury), as in severe shock, urine formation ceases. Thus urine formation begins with the ultrafiltration of a large volume of blood plasma from the glomerular capillaries into the capsular space, colloids such as proteins being held back while crystalloids pass through. The only difference between plasma and the initial filtrate is the absence from the latter of molecules above a borderline size. The filtrate contains all the substances present in the blood, and in the same concentration, except the colloids and fats.

Table 19.1 Relative composition of plasma and urine in normal men

	Plasma g/100 ml	Urine g/100 ml	Concentration g/100 ml
Water	90–93	95	–
Proteins and other colloids	7–8.5	–	–
Urea	0.03	2	× 60
Uric acid	0.002	0.03	× 15
Glucose	0.1	–	–
Creatinine	0.001	0.1	× 100
Sodium	0.32	0.6	× 2
Potassium	0.02	0.15	× 7
Calcium	0.01	0.015	× 1.5
Magnesium	0.0025	0.01	× 4
Chloride	0.37	0.6	× 2
Phosphate	0.003	0.12	× 40
Sulphate	0.003	0.18	× 60
Ammonia	0.0001	0.05	× 500

TUBULAR ABSORPTION AND CONCENTRATION

Urinary secretion is not, however, a matter of simple filtration with subsequent concentration by reabsorption of water back into the blood in the capillaries surrounding the tubules. There is also a *selective* action in the tubule cells which accounts for the fact that, while there is 60 times as much urea in the urine as in the blood, there is only twice as much chloride. The filtrate from the glomeruli enters the tubules, where most of the water and some electrolytes are reabsorbed into the bloodstream, while wastes like urea are selectively retained in the urine.

80–85% of the water and sodium in the filtrate is reabsorbed in the tubules. For each constituent of the blood, there is a threshold concentration which must be exceeded before it appears in the urine. Thus sugar is normally never excreted because its threshold is rarely exceeded, save in diabetes; in other words, all the glucose arriving in the filtrate is reabsorbed. On the other hand, reabsorption of chlorides is much less, and so they are normal constituents of urine. Further, the urine may be acid or alkaline and so compensate for excessive intake of alkalis in the food or overproduction of acids in the body.

Renal clearance relates to the volume of plasma from which a particular substance is completely removed by the kidneys in one minute. Clearance is given in the formula:

$$RC = UV/P$$

where U is the concentration of the substance in the urine in mg/ml, V is the rate of urine formation in ml/min and P is the concentration of the substance in the plasma.

THE PITUITARY AND DIURESIS

The final concentration of the urine by water reabsorption into the circulation occurs in the main collecting tubules of the kidney. This is controlled by an *antidiuretic hormone* (ADH) secreted by the posterior lobe of the pituitary gland (Chapter 21). This hormone increases the permeability of the tubule walls to water. Its secretion is controlled by a feedback mechanism sensitive to the osmotic pressure of plasma and tissue fluids. This is mediated by osmoreceptors at the base of the brain. At normal blood osmolarity, there is a steady receptor discharge and a steady ADH output. But if the plasma becomes hypotonic from ingestion of excess sodium chloride, receptor discharge increases, ADH output rises, the kidney tubule walls become more permeable, and a smaller volume of more concentrated urine is passed, i.e. there is an *antidiuretic effect*, the object of which is to retain water in the system to dilute the excess salt. On the other hand, excess ingestion of water dilutes the body fluids, ADH secretion is reduced, the tubule walls become less permeable and there is a *diuresis*, i.e. the passage of a large amount of pale dilute urine. The mechanism now functions to rid the body of excess water.

KIDNEY HORMONES

As well as its excretory functions the kidneys secrete two hormones:

▸ *renin*, which helps regulate blood pressure

▸ *erythropoietin*, which stimulates red blood cell formation in the bone marrow.

They also produce *nitric oxide*, which is a potent vasodilator, and other agents affecting the calibre of blood vessels generally.

THE BLADDER AND MICTURITION

Urine is propelled along the ureters by waves of contraction in their muscular walls. It is discharged from the ureteric orifices into the bladder in intermittent jets.

The bladder, like other hollow organs, is a muscular sac whose outlet is normally closed by a tight ring of muscle, or *sphincter*. Bladder wall and sphincter must be reciprocal in action. In emptying, the bladder contracts and the sphincter relaxes to allow the outflow of urine, and in the resting phase the bladder is relaxed to allow gradual distension as urine collects, while the sphincter remains closed. *Micturition* is the expulsion of urine from the bladder along the urethra. The control of the bladder in this respect is both voluntary and involuntary. Basically, the organ is self-regulating and empties itself once internal pressure reaches a certain level. The *sympathetic* and *parasympathetic* parts of the independent autonomic nervous system are the regulators. The parasympathetic nerves empty it by contracting its walls and relaxing its sphincter; the sympathetic allows it to fill by a reverse action.

But there is also a voluntary control which can modify or override the basic reflexes. There is a second sphincter around the urethra itself and this is under the control of the will, so that micturition is normally deliberate and aided by voluntary contraction of the abdominal muscles. But once initiated, the process continues under automatic control. A paraplegic person has a perfectly functioning 'autonomous bladder', though it is not completely emptied.

Test yourself

1 How much of a kidney is required to carry out the excretory function essential to life?
   a   Half
   b   One-third
   c   Two-thirds
   d   Three-quarters

2 Identify a role that the kidney does not carry out.
   a   Regulating water content
   b   Regulating pH
   c   Regulating osmotic pressure
   d   Regulating blood glucose

3 What makes urine frothy when shaken?
   a   Mucus
   b   Urea
   c   Bile salts
   d   Red blood cells

4 What makes the ammonia found in urine?
   a   The breakdown of red blood cells
   b   The breakdown of protein
   c   The breakdown of minerals
   d   The breakdown of carbohydrate

5 Which condition could lead to glucose in a person's urine?
   a   Pregnancy
   b   Malaria
   c   Liver disease
   d   Diabetes

6 Where are antidiuretic hormones secreted from?
   a   Pituitary gland
   b   Pancreas
   c   Medulla
   d   Liver

7 When would more antidiuretic hormone be released?
   a   After a person has eaten a lot of salty foods
   b   After a person has drunk lots of water
   c   After a person has eaten a lot of sweet foods
   d   After a person has eaten a lot of fruit

8 Identify a hormone released by the kidney.
   a   Glucagon
   b   Insulin
   c   Renin
   d   Adrenaline

9 What is the equation for renal clearance?
   a   RC = UP/V
   b   RC = VP/U
   c   RC = V/UP
   d   RC = UV/P

10 What agent released by the kidneys produces vasodilation?
   a   Lactic acid
   b   Carbonic acid
   c   Nitric oxide
   d   Potassium

Nervous system and sense organs

In this chapter you will learn about:

▶ *the basic structure of the sense organs*

▶ *the neurone and nervous transmission*

▶ *the anatomy of the central nervous system*

▶ *the cerebrospinal system*

▶ *the spinal cord*

▶ *the brain*

▶ *the autonomic nervous sytem*

▶ *special sense organs, including the skin, eye, ear and nose.*

All protoplasm is *excitable* and *conducts* excitation, but especially nervous tissue. Nerve cells are particularly sensitive, and their fibres specialize in the transmission of impulses. The nervous system is a network throughout the body, with a two-way connection with central control and permitting a coordinated response to any stimuli from outside or within.

The main parts of the nervous system are:

▶ the *central nervous system* (brain and spinal cord)

▶ the *peripheral nervous system*, long bundles of fibres attached to central cells.

The receptor side of this system conveys information from outside by *afferent* impulses along nerve fibres to the control centres, but it also provides information as to the internal state of the body. Some of these messages reach consciousness, but others, notably those from within, remain largely unconscious. The *effector* side of the system carries *efferent* messages to the effector organs – muscles and glands – and this response is meant to deal with the situation provoking the original sensory stimuli. From the physiological viewpoint, the brain is only the centre of a glorifed reflex arc, to which consciousness has been added as an incidental by-product.

Basic structure

The basic elements of the nervous system are:

1 the nerve cells, or *neurones*

2 the supporting cells within the central nervous system, or *neuroglia*

3 *connective tissue*, the *membranes* around the brain and cord and the *sheaths* of the nerve fibres, though there is no connective tissue in the actual brain or cord substance.

THE NEURONE (FIGURE 20.1)

Each nerve cell has several branching processes, or *dendrites*, interlocking with those of adjacent cells. One is often elongated as the main *axon* for transmission of stimuli. The axon may be enormously long, traversing the lengths of the limbs or spinal cord. Cell, dendrites and axon constitute the nerve unit, or *neurone*, and the nervous system is built up of millions of such interrelated units. They do not divide and have a high metabolic rate demanding abundant oxygen and glucose. They are progressively lost in the brain with age and cannot be replaced, but may survive a lifetime.

Neurones communicate by *synapses* between the axons of one neurone and the dendrites or cell bodies of others. Transmission may be:

▶ *electrical*, through protein channels allowing direct passage of ions from one neurone to another

▶ *chemical*, in which electrically excited neurotransmitters diffuse across a narrow cleft between cells, where their signals are reconverted to electrical changes.

The main neurotransmitters are classically acetylcholine, adrenaline (and noradrenaline), dopamine and serotonin, but recent work has shown the importance of *nitric oxide* as a signalling gas passing rapidly into cells and binding briefly to iron-containing enzymes before disappearing.

There are two kinds of peripheral nerve fibres: *myelinated* and *unmyelinated*. The large white or myelinated fibres have a fatty sheath or *neurilemma* formed by investing Schwann cells,

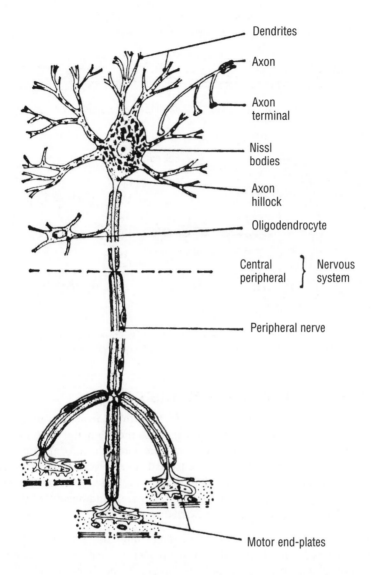

Dendrites

Axon

Axon
terminal

Nissl
bodies

Axon
hillock

Oligodendrocyte

Central
peripheral } Nervous
system

Peripheral nerve

Motor end-plates

Figure 20.1 Diagram of an anterior cell from the spinal cord (from *A Companion to Medical Studies*, Vol. 1)

which is indented at intervals by the nodes of Ranvier (Figure 20.2). These fibres are also found in the *white matter* of the central nervous system and conduct impulses rapidly. The unmyelinated fibres are finer and occur in the autonomic nervous system and the *grey matter* of the brain and cord and conduct impulses slowly. The nerve bundles of a peripheral nerve contain both kinds of fibres; they are bound together by a connective tissue sheath – the *perineurium* – and the sheath of the entire nerve is the *epineurium* (see Figure 20.3).

Damage to nerve cells is irreparable as they are incapable of reproduction. However, the function lost in destroyed areas in the cerebral cortex may be taken over by other parts of the brain. There is some hope that cell transplants may be useful here. Injured fibres may grow again, but only in the peripheral nerves after careful surgical repair.

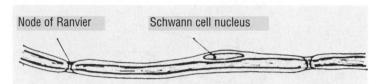

Figure 20.2 Internodal segment of a nerve (from *A Companion to Medical Studies*, Vol. 1)

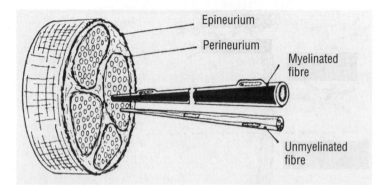

Figure 20.3 Diagram of the structure of a peripheral nerve containing myelinated and unmyelinated nerve fibres, to show the neural sheaths (from *A Companion to Medical Studies*, Vol. 1)

The nature of nervous transmission

The nerve fibres only *conduct* stimuli; those passing from the central nervous system to muscles or glands or those from skin and sense organs to the brain and cord. Fibres normally conduct only in one direction. Nerves are like telephone wires; the information conveyed largely depends on who happens to be connected at either end. We associate the optic nerve with sight and the olfactory nerve with smell, but they serve these particular functions only because they link the appropriate sense organ with the corresponding area of the brain. The impulses both transmit are identical to those carried by all nerves.

Nerve fibres are readily stimulated experimentally and stimulation sets up a wave of depolarization, or *action potential*, travelling at many metres a second. The action potential is a brief reversal of cell membrane potential with an amplitude change of about 100 mV. This wave is not the nerve impulse itself, but the index of an underlying disturbance, a sudden change in local membrane permeability to ions, particularly sodium and potassium, i.e. there is an *electrochemical gradient*.

Nervous action is a stream of small separate impulses, repeated at very short intervals. There is no continuous flow of excitation, as each impulse is followed by a short *refractory period* during which the nerve will not transmit a stimulus.

A single fibre exhibits the *all-or-none* phenomenon, i.e. if it is adequately stimulated, it will convey an impulse of fixed magnitude. However, smaller stimuli do not produce any impulses and larger stimuli do not produce a larger impulse. The large sheathed motor and sensory fibres of the central nervous system carry impulses at 90 m/s; in the finer fibres of the autonomic system the speed is only some 1.5–15 m/s. These differences are related to the need for an immediate response to an external situation, whereas the regulation of the internal organs is a more leisurely process. Like the body tissues, nervous tissue undergoes metabolic activity, using oxygen and generating carbon dioxide with the production of heat. Though the energy amounts are minute compared with those of muscle activity, nervous tissue, particularly the brain cells, cannot tolerate oxygen deprivation for more than a very few minutes.

In a single cell and its fibres the impulse is conducted uninterruptedly, but there is an inevitable break when it is *relayed* to another cell. Such relays, or *synapses*, are an integral part of the nervous system. Motor impulses from the brain are relayed by other cells in the spinal cord, whose fibres run to the muscles, and sensory impulses from the skin are relayed by cord cells before passing to the brain. The incoming fibre ends in a network of branches which embrace those of the relay cell but there is no physical continuity. Because of this gap messages are slightly delayed at the synapse.

The bridging of the synapse gap is effected by chemical means; the nerve endings liberate chemical substances which stimulate the relay cell to start a fresh impulse along its own fibre. A similar gap must be crossed when a motor fibre ends in a muscle, and this is also achieved by liberation of a chemical agent, acetylcholine. Though nerve fibres can conduct in either direction, transmission at the synapse is strictly one-way; hence the directional flow in the central nervous system. The same applies at the neuromuscular junction.

The *chemical mediators* are largely acetylcholine and noradrenaline. These are very potent and closely associated with the autonomic nervous system, whose ultimate fibres to glands and viscera are labelled *cholinergic* or *adrenergic* as the case may be (see below). We shall see that the autonomic system uses chemical agents to play a very large part in regulating internal organs and blood vessels, and that certain endocrine glands (e.g. the adrenals) can secrete the same substances directly into the blood as hormones to secure a rapid response in emergency.

Anatomy of the central nervous system

The brain and spinal cord, enclosed in the cranium and spinal canal, are continuous at the foramen magnum at the skull base. The twelve pairs of cranial nerves arising from the brain and the 32 pairs of spinal nerves arising from the cord constitute the *peripheral* nervous system. Together, they form the *cerebrospinal* or *voluntary* nervous system, mainly concerned with control and sensation of the somatic structures of the body wall – skin, muscle, bones and joints – though many of its activities are unconscious.

The semi-independent *autonomic* or *vegetative nervous system* deals with the automatic functioning of the splanchnic structures – viscera, glands and vessels. However, the two systems are closely interconnected.

Cerebrospinal system

MEMBRANES AND CEREBROSPINAL FLUID

The brain and spinal cord have three enveloping membranes, which are prolonged as sheaths along the issuing nerve roots. The outermost layer is the *dura mater*, a tough loosely applied protective envelope. In the cranium it also forms the inner lining periosteum of the skull bones. The innermost layer is the *pia mater*, a fine membrane closely applied to the brain and cord, filling every cleft and crevice, and carrying with it the fine blood vessels. Intermediate is the *arachnoid layer*; this fits loosely within the dura but there is a *subarachnoid space* separating it from the pia, filled with *cerebrospinal fluid* and traversed by spidery strands of connective tissue (Figure 20.5).

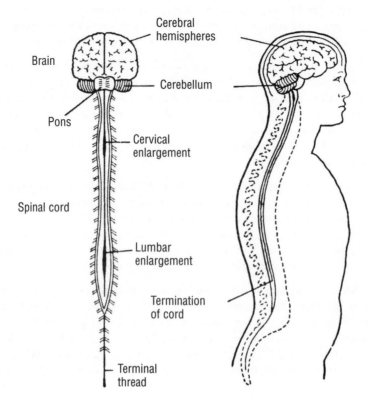

Figure 20.4 The central nervous system

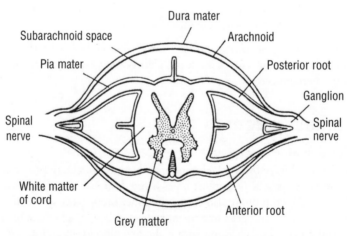

Figure 20.5 Cross-section of the spinal cord and its membranes (after *Gray*)

The cerebral and spinal membranes are continuous at the foramen magnum. The fluid bathes the outer surfaces of brain and cord. The fluid also occupies the hollow chambers or *ventricles* of the brain, which communicate with the central canal of the cord. It is secreted by the vascular *choroid plexus* lining the ventricles, circulates inside the brain and cord, escapes through the roof of the hindbrain into the subarachnoid space, and is reabsorbed into the bloodstream via the venous sinuses of the cranium. There is a continuous circulation of fluid from the blood vessels in the ventricles, in and around the brain and cord, and back into the bloodstream.

SPINAL CORD

The spinal cord is elongated and cylindrical with two swellings, the *cervical* and *lumbar* enlargements, which are the origins of the roots of the brachial and lumbar nerve plexuses for the upper and lower limbs.

In the foetus the cord occupies the entire length of the spinal canal, but it fails to keep pace with growth and in the adult ends at the first lumbar vertebra in a conical extremity. Since the nerve roots must still emerge from the intervetebral formina at the correct levels, they have to run more and more obliquely downward and outward as they arise lower in the cord, so the spinal canal below the termination of the cord is filled wth a mass of roots – the *cauda equina* or 'horse's tail' – descending to their lumbar and sacral exits.

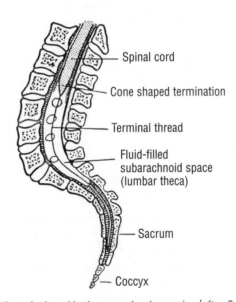

Spinal cord

Cone shaped termination

Terminal thread

Fluid-filled
subarachnoid space
(lumbar theca)

Sacrum

Coccyx

Figure 20.6 The termination of the spinal cord in the upper lumbar region (after *Gray*)

Cross-section of the cord shows the inner H-shaped arrangement of *grey matter*, composed of nerve cells, and the surrounding *white matter* – descending and ascending tracts of nerve fibres. In the very centre is a minute canal. In the midline a cleft anteriorly and a fissure posteriorly divide the cord into symmetric right and left halves.

On each side the grey matter projects in front and behind as the anterior and posterior horns. The *anterior horn* contains the motor nerve cells, whose fibres are destined for stimulation of muscles, and some of these leave the cord in a bundle as the anterior or motor nerve root at each intervertebral foramen.

The *posterior horn* contains sensory cells; fibres enter it from the posterior or sensory nerve root and are mediated by a group of cells forming a knob or *ganglion* on this root just outside the cord (see Figure 20.7).

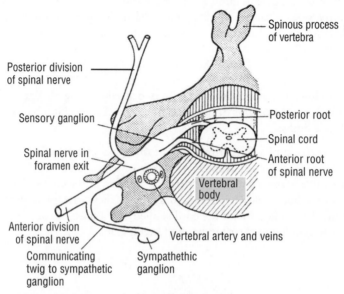

Figure 20.7 Cross-section showing the spinal cord, its issuing roots and the appropriate spinal nerve, as they lie in relation to the bony spine; the section is in the cervical region

Note the communicating twig to the sympathetic ganglion (after *Gray*)

SPINAL NERVES

The two roots on each side join just beyond the sensory ganglion to form the spinal nerve proper, their junction lying within the intervertebral foramen before the nerve leaves the spine. Thus the spinal nerve contains sensory and motor fibres, and these are not separated again until, in the nerve branches in limb or trunk, specific motor and sensory twigs are given off to the muscles and skin.

The simplest pattern is in the thoracic region, the nerve on each side encircling the chest in its appropriate intercostal space with no connection with its neighbour above or below. But the lower cervical and lumbar and sacral nerves form complex networks as soon as they have left the spinal column. From these plexuses emerge the peripheral nerves of the limbs: the median, radial and ulnar in the arm, sciatic and femoral in the lower limb; these are usually mixed motor and sensory, though some are almost entirely one or the other.

The cross-section of the cord is much the same at all levels, i.e. the cord consists of a number of identical (but not externally demarcated) *segments*, to each of which is attached a pair of spinal nerves. This is a relic of the primitive segmentation which lingers on in the repetitive arrangement of ribs and vertebrae, and it is important because each cord segment controls the sensation of a particular area of skin, or *dermatome*, and the movement of a particular group of muscles. Thus the sensation of the little finger is always supplied by the first thoracic segment of the cord, the muscles straightening the knee always derive their motor fibres from the third and fourth lumbar segments, and so on. This is of great value in locating the precise site of injury or disease of the central nervous system.

THE REFLEX ARC

The spinal cord provides a relatively simple means for performing essential actions without the intervention of the brain. Irritation of the skin sends a sensory impulse through the posterior root to the posterior horn of the appropriate segment. This is relayed to the anterior horn cells of the same or related segments and becomes a motor impulse travelling out through the anterior root to activate muscles which withdraw the affected part. This is the reflex arc at its simplest, and is independent of the brain, though the brain is aware of its results and can modify them. Another simple arc is involved in the *tendon-jerk*. The simplest example is the knee-jerk; when one leg is crossed over the other a tap on the patellar tendon causes the quadriceps muscle to straighten the knee. There are many similar reflexes, based on stimulation of stretch receptors in the muscle.

More complex but basically similar are the reflexes of micturition, defaecation, orgasm and childbirth, all controlled by a particular segment or group of segments. All these activities persist even when the cord-brain link has been severed and the body is paralysed as regards conscious sensation and voluntary movement.

THE TRACTS OF THE CORD (FIGURE 20.11)

The grey matter consists mainly of cells. The white matter is made of fibres running longitudinally in cable fashion, grouped in a constant pattern of individual tracts, some sensory, some motor. A few tracts run for several segments only, but most run between cord and brain and link the anterior or posterior horn cells with the appropriate centres of the brain. Thus, from the motor areas of the cerebral cortex concerned with arm movement, fibres run to the anterior horn cells of the seventh cervical cord segment, and from these other fibres leave to join the brachial plexus, enter the median nerve and activate the flexors of the wrist. Sensory impulses from the knee travel up the femoral nerve, through the lumbar plexus to the posterior horn cells of the fourth lumbar segment, and are relayed up the long sensory tracts to the sensory portion of the cerebral cortex.

In other cases, other parts of the brain are the termini – cerebellum or midbrain – but an invariable principle is that of crossing or *decussation*. At some point, usually in the brainstem, the tracts cross from one side to the other, so that the sensory and motor regions of the brain control the opposite halves of the body.

THE BRAIN (CEREBRUM)

The brain is the greatly elaborated upper end of the cerebrospinal axis. It has the same membranes as the cord, it almost entirely fills the cranial cavity, and it makes indentations on the inner aspect of the cranial bones. The grey matter is now on the surface and the white fibre tracts within. The main parts of the brain, from above down, are the *forebrain*, the *midbrain* and the *hindbrain*.

▶ The forebrain

The forebrain is the great overhanging pair of *cerebral hemispheres* – the bulk of the organ – symmetric rounded masses of convoluted nervous tissue which hide the lower parts of the brain viewed from above.

The two hemispheres are separated by a deep fissure, with a partition of dura mater, or *falx*, but are connected at the base of this cleft by a great bridge of fibres, the *corpus callosum*. The superficial grey matter, or cerebral cortex, has an enormous area owing to its intricate convolutions, which produce a number of ridges, or *gyri*, separated by valleys or *sulci*.

Each hemisphere is composed of several *lobes* which are not clearly demarcated. In side view, the *frontal lobe* is at the anterior pole of the hemisphere, in the anterior cranial fossa. At the back is the *occipital lobe,* in the posterior cranial fossa. The *temporal lobe* lies behind the frontal and projects below, occupying the middle cranial fossa, and the *parietal lobe* is an ill-defined area between frontal and occipital regions above the temporal lobe.

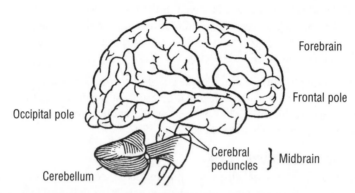

Figure 20.8 The brain in side view; diagrammatic (after *Gray*)

The main motor and sensory areas are midway between the poles, and the body is represented in them in inverted fashion, i.e. the areas for the legs are uppermost, those for the head below. There is a vaguely localized centre for appreciation of the meaning of words in the temporal lobe of the left side, on the right side in left-handed people All these functional areas are concerned with the opposite half of the body, though the visual area of the occipital cortex is concerned not so much with the opposite eye as with gathering all the impulses from the opposite field of vision that have entered both eyes.

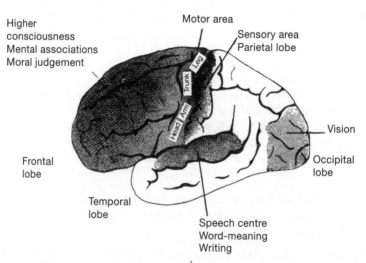

Figure 20.9 Functional areas in the left cerebral hemisphere; the right side is identical, save for the absence of a speech centre

Much of the cortex is assigned no particular function. These are the psychic or association areas, concerned with correlation of data and the higher levels of personality, particularly in the frontal lobes.

Cross-section of the forebrain (Figure 20.10) shows the pattern of sulci and gyri, the contrast of grey and white matter, and the cleft between the hemispheres. On each side is the hollow *ventricle* in the depth of the hemisphere, filled with cerebrospinal fluid, and great masses of grey matter beside the ventricles, the *basal ganglia*, one of which – the *thalamus* – is related to the emotions.

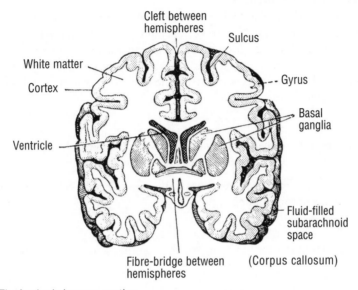

Figure 20.10 The forebrain in cross-section

▶ The midbrain

The midbrain is situated at the base of the hemispheres and consists partly of a squat *peduncle* on each side, a pillar supporting the corresponding hemisphere and carrying fibres to and from it. The roof of the midbrain contains the nuclei of the third cranial nerve, which controls movement of the eyeballs.

The *pituitary body* projects from the undersurface of the midbrain in front of the peduncles (Figure 21.2) and the *pineal body* from its roof.

▶ The hindbrain

The hindbrain is best seen on the underside of the organ (Figures 20.8, 20.12). It includes:

▶ the *medulla oblongata*, the bulbous upward prolongation of the spinal cord, the lowest part of the brain or *brainstem*, containing the nerve centres for heartbeat and respiration

▶ the *cerebellum*, a pair of rounded hemispheres, finely convoluted, occupying the deepest part of the posterior cranial fossa

▶ the *pons*, a broad bridge of fibres connecting the cerebellar hemispheres.

The hindbrain is the most primitive part of the organ and is concerned with the most basic activities of the body. It is independent of conscious control and continues to function in sleep, unconsciousness and deep anaesthesia. The medulla's vital centres regulate heartbeat, respiration and blood pressure and remain active until the last moment before death. The cerebellum deals with the regulation of muscle tone and posture in response to the sensory input of position and tension from joints and tendons, also of information on the body's position in space from the semicircular canals of the inner ear. It governs the postural mechanisms, the righting reflexes that preserve upward carriage of the head.

FIBRE TRACTS OF THE BRAIN AND CORD

A series of cell stations is situated at higher and lower levels in the grey matter of the cerebral cortex, basal ganglia, midbrain, medulla and cerebellum. To and from these, great tracts of nerve fibres ascend and descend within the white matter, the axial core of medulla and hindbrain. These tracts were encountered in cross-section of the cord, and may be classified as *motor* (descending) and *sensory* (ascending). At some level, usually in the hindbrain, they all cross to the opposite side.

The chief motor tracts are:

▶ *pyramidal*, running from the cerebral motor cortex to the anterior horn cells of the cord at all levels, thence relayed to the motor roots, concerned with voluntary motion

▶ *cerebellospinal*, running from the cerebellum to the anterior horn cells, concerned with automatic regulation of tone and posture.

The chief sensory tracts are:

▶ the *posterior bundles* of the cord carrying touch and other sensation to the cerebral sensory cortex

▶ the *anterior* and *lateral bundles* carrying pain and temperature sensation to the cortex

▶ the *spinocerebellar bundle* carrying postural sensation to the cerebellum.

The sensory bundles relay impulses entering the cord via the posterior sensory roots and posterior horn cells. Since they acquire an increasing influx as they ascend, and since the motor tracts shrink as they descend because they shed fibres to successive segments, the cord as a whole tapers from above downwards.

THE CEREBRAL VENTRICLES

The brain is hollow and traversed by channels for the cerebrospinal fluid. In each cerebral hemisphere is a large *lateral ventricle*, and between and below these is a small *third ventricle*, from the floor of which the pituitary body is suspended by its stalk. These communicate with a tiny channel, the *aqueduct*, which leads back through the midbrain to the *fourth ventricle* of the medulla. This is continuous with the central canal of the spinal cord, and from its roof the fluid escapes to bathe the outer surface of the brain and cord in the subarachnoid space. Growing from the linings of the lateral and fourth ventricles is a very vascular *choroid plexus*, which secretes the fluid; any block to its free flow within the brain results in a damming-up which balloons the brain to produce the condition of *hydrocephalus* or 'water on the brain'.

The cerebrospinal fluid cushions the brain and cord against concussion or sudden changes of posture. It is normally maintained at a constant pressure which is related to the blood pressure

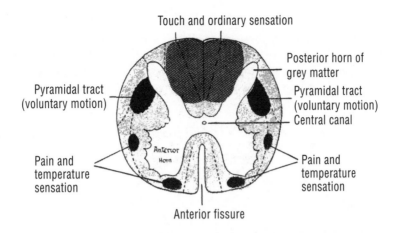

Figure 20.11 Cross-section of spinal cord, showing the important nerve tracts

in the great veins, and is raised when venous pressure is increased by coughing or straining. Any disease process or tumour within the unyielding cranium increases the cerebrospinal fluid pressure as it expands, causing headache and vomiting.

The composition of the fluid and its contained cells may be affected by disease, so examination of the fluid is often useful in diagnosis. While the cord ends at the first lumbar vertebra, its membranes continue as far as the sacrum. The lumbar spine thus encloses a space or *theca*, filled with cerebrospinal fluid and nerve roots, and this may be entered by a long needle inserted between the spinous processes: the procedure of *lumbar puncture* or *spinal tap*.

VESSELS AND NERVES OF THE BRAIN

The main *arterial* supply of the brain is derived from the two internal carotids, which enter the skull base. Together with the vertebral arteries, which have entered the foramen magnum on either side of the cord, they form a vascular circle around the stalk of the pituitary body. From this, *anterior*, *middle* and *posterior cerebral arteries* are given off to the frontal, parietotemporal and occipital regions of the brain. The hindbrain is supplied by the vertebral artery.

The main *veins* form *sinuses* which run between the layers of the dura mater and drain into the internal jugular. The *superior sagittal sinus* of the vault, which runs the length of the brain from front to back between the hemispheres, also drains off the cerebrospinal fluid, which is constantly being reabsorbed.

The brain substance has no nerves and can be handled without reaction, even in the conscious patient; but its covering membranes are very sensitive.

CRANIAL NERVES

In certain parts of the brain there are collections of nerve cells forming the *nuclei* from which spring the twelve pairs of cranial nerves, best seen on the underside. These are concerned with movement and sensation in the head and face, including the eye, ear and nose; but one (the eleventh) controls certain neck muscles and another (the tenth or vagus) is distributed to the internal organs of the neck, chest and abdomen. They may be listed from above downwards:

▶ **Cerebrum**

(I) The first or olfactory nerves are those of smell. On each side a group of some twenty rootlets ascends from the nose through perforations in the floor of the anterior cranial fossa, and ends in the olfactory bulb under the frontal lobe. From this an olfactory stalk or tract relays the impulses to the cerebral hemisphere. These nerves are purely sensory.

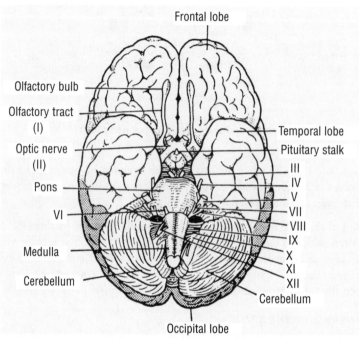

Figure 20.12 The underside of the brain, showing the attachments of the twelve pairs of cranial nerves

(II) The second or *optic nerve* is concerned with sight. Each carries fibres from the nerve cells of the retina and leaves the back of the orbit as the stalk of the eyeball through the optic foramen to enter the middle cranial fossa. The two nerves form an X-shaped crossing or *chiasma* under the frontal lobes, in front of the pituitary stalk. The two hind limbs of the X are the *optic tracts*. The effect of this is that fibres from the outer half of each retina remain in the optic tract of the same side, while those from the inner halves cross into the opposite tract. Thus the right tract carries all the fibres from the right halves of both retinae and therefore all the sensation from the left half of the visual field, and vice versa for the left tract. This separation is essential to maintain one-sided representation of *function* in each half of the brain. The eye is a one-sided organ, but sees on both sides of the body, and what is needed for coordination is that each side of the visual field should be represented as such in the cerebral cortex of the occipital lobe. The impulses are relayed in the visual area there and are connected with the nucleus of the third nerve, which controls eye movements. The optic nerve is purely sensory.

▶ **Midbrain**

(III) The third or *oculomotor nerve* controls four of the six small muscles moving the eyeball, and is purely motor.

(IV) The fourth or *trochlear nerve* controls another of the ocular muscles and is purely motor.

▶ **Hindbrain**

(V) The fifth or *trigeminal nerve* has a large *sensory root* divided into three branches:

   a *ophthalmic*: this conveys sensation from within the orbit other than sight, and sensation from the forehead and front of the scalp

   b *maxillary*: this conveys sensation from the upper jaw and teeth and overlying skin

   c *mandibular*: this conveys sensation from the lower jaw and teeth and overlying skin, and ordinary sensation from the front of the mouth and tongue.

The mandibular division enters a canal in the lower jaw. The ophthalmic division enters the back of the orbit. The maxillary branch runs in a canal between the orbital floor and the roof of the maxilla to emerge on the front of the cheek.

The fifth nerve also has a small *motor root*, concerned with the muscles of mastication. Thus it is a mixed motor and sensory nerve.

(VI) The sixth or *pathetic nerve* controls the last of the ocular muscles. It is purely motor.

(VII) The seventh or *facial nerve* supplies the facial muscles; it also activates the secretion of some of the salivary glands and conveys taste sensation from the front of the tongue. It leaves the posterior fossa of the skull by entering the internal auditory meatus on the inner aspect of the temporal bone with the eighth nerve. After a winding course it emerges in front of the mastoid process and enters the parotid gland, in which it divides into twigs supplying the facial muscles and the platysma in the neck. It is a mixed sensory and motor nerve.

(VIII) The eighth or *auditory nerve* connects the inner ear, deep in the temporal bone, with the hindbrain; it serves two distinct functions.

   a The *acoustic* portion carries sensations of sound and pitch from the organ of hearing, the cochlea, which are relayed to the auditory part of the cerebral cortex.

   b The *vestibular* portion carries postural sensation from the semicircular canals of the inner ear to the cerebellum, providing information about position.

(IX) The ninth or *glossopharyngeal nerve* is mainly sensory. It conveys sensation from the pharynx, tonsil and back of tongue, including taste; but it also activates parotid secretion and some small muscles of the pharynx.

(X) The tenth nerve, the *vagus*, is very important, extending through the neck and thorax (see Chapters 9 and 10) to the abdomen as an essential component of the autonomic system controlling the viscera. Its main branches are:

   a in the *neck*, for movements of the pharynx and larynx, and sensation from their linings

   b in the *chest*, where it forms the cardiac and pulmonary plexuses, supplies the muscle of the oesophagus and bronchi and their associated glands and sensation from their linings

   c in the *abdomen*, where it supplies the muscle, glands and mucous membranes of the stomach and duodenum, and sends twigs to the liver, spleen and kidneys.

(XI) The eleventh or *accessory nerve* is purely motor and supplies the muscle of soft palate and pharynx and the sternomastoid and trapezius muscles in the neck.

The ninth, tenth and eleventh nerves all arise from the medulla and leave the skull through the jugular foramen with the internal jugular vein.

(XII) The twelfth or *hypoglossal nerve* controls the muscles of the tongue and floor of the mouth.

The fibres of motor cranial nerves and those of the motor roots of the cord arise from their cells in the cerebral nuclei or anterior horns and pass without a break to their destined muscles. But sensory fibres entering from the skin or other surfaces end in a group of cells just outside the brain or cord which form *ganglia*, knobby swellings on the posterior spinal roots or on the sensory cranial nerves at the skull base. All sensory nerves have a ganglion and it is the axons of the ganglion cells, not the original sensory fibres, that enter the cerebrospinal axis, i.e. motor impulses are direct, sensory impulses are mediated. The outgoing motor pathways are known as *efferent*, the incoming sensory fibres are *afferent*.

Autonomic nervous system

SYMPATHETIC AND PARASYMPATHETIC (FIGURE 20.13)
The autonomic nervous system is distributed to the muscle coat of the blood vessels and the smooth muscle and glands of the viscera. It controls their movements and secretions and monitors their distension. All its fibres are ultimately derived from the cerebrospinal axis itself, and it is susceptible to modification from the centre, not by an act of will but through the emotions.

▶ Anatomical arrangements

The *sympathetic system* is a chain of ganglia on each side of the spinal column from atlas to coccyx, where the two meet and fuse in a single ganglion. This chain lies on the sides of the vertebral bodies at the back of the abdominal and thoracic cavities, where there is one ganglion to each segment and spinal nerve. In the neck there are only three. The outflow from the central nervous system to the sympathetic is by a small communicating twig from each spinal nerve in the thoracic and upper lumbar regions to the corresponding sympathetic ganglion (Figure 20.7). From this ganglion the peripheral sympathetic system develops in two ways:

1 by a returning twig to each spinal nerve, the sympathetic fibres travelling out with that nerve to the somatic structures of the limbs and body wall

2 by the formation of *plexuses*, e.g. the cardiac and pulmonary plexuses, branches of which form a network around the vessels to the heart and lungs; also the *splanchnic nerves* run down from the lower thoracic chain through the diaphragm to form the *coeliac (solar) plexus* around the upper part of the abdominal aorta from which branches go to the viscera.

The *parasympathetic system* is a less distinct structure, as its fibres are contained within certain cranial and sacral nerves at either end of the cerebrospinal axis: the *craniosacral outflow*. The former supply the glands and vessels of head and eye, the latter the bladder, rectum and genital organs. But the *vagus* – the tenth cranial nerve – is the great parasympathetic nerve of the body, wandering through neck, chest and abdomen to supply the viscera as a whole.

There is a major difference between the autonomic and central nervous systems. In the latter the cell bodies of the neurones lie *within* the system; only fibres lie outside the brain and spinal

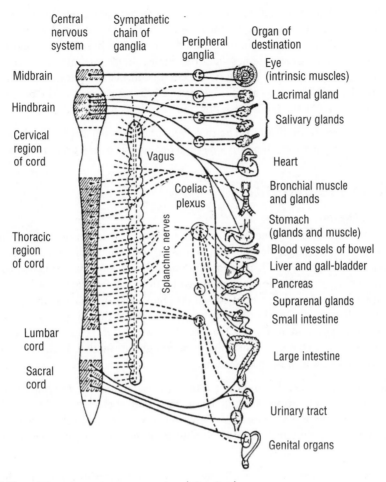

Central nervous system

Sympathetic chain of ganglia

Peripheral ganglia

Organ of destination

Midbrain

Hindbrain

Cervical region of cord

Vagus

Thoracic region of cord

Coeliac plexus

Splanchnic nerves

Lumbar cord

Sacral cord

Eye (intrinsic muscles)

Lacrimal gland

Salivary glands

Heart

Bronchial muscle and glands

Stomach (glands and muscle)

Blood vessels of bowel

Liver and gall-bladder

Pancreas

Suprarenal glands

Small intestine

Large intestine

Urinary tract

Genital organs

Figure 20.13 Plan of the autonomic nervous system (after *Gray*)

cord. In the autonomic, there are *peripheral ganglia* consisting of cell stations with synapses. In the sympathetic these are located close to the vertebral bodies; in the parasympathetic they are near or in the organs. Fibres in the autonomic system are either preganglionic or postganglionic. The parasympathetic has long preganglionic fibres in a spinal nerve before synapsing with ganglionic cells close to the organ they supply, so their postganglionic fibres are very short. Conversely, the sympathetic outflow has short preganglionic fibres but their postganglionic course is often very long.

Chemical transmission in the autonomic system takes place:

▶ at the ganglionic synapse

▶ where the postganglionic fibres enter the target organ.

At the former site both sympathetic and parasympathetic act by liberating *acetylcholine*. At the peripheral site the sympathetic liberates acetylcholine where sweat glands and skeletal blood vessels are concerned, but noradrenaline at the heart, visceral muscle, glands and internal vessels; the parasympathetic again liberates acetylcholine. The actions of these agents are completely antagonistic. Adrenaline is the chief sympathetic transmitter at the end-organs.

The effects of sympathetic stimulation are those of bodily activity in relation to fear, fight or flight. Aided by adrenaline secreted by the adrenal glands, it constricts the vessels of the skin, raising the blood pressure and shunting blood to heart and brain, speeds and strengthens the heartbeat, dries glandular secretion, dilates the pupils, stands the hair on end and initiates sweating, and relaxes the walls of the hollow viscera. In contrast the parasympathetic promotes relaxed constructive activities in tranquillity, as after a heavy meal, dilates the peripheral vessels, slows the heart and lowers the blood pressure, and excites secretion and peristalsis.

However, this is an oversimplification. The two parts of the system always work in tandem to control the internal organs, even if at any one time one or other may predominate.

Special sense organs

SENSATION IN GENERAL

Information from the muscles, joints and internal organs is constantly arriving at the brain; but, except pain, these messages do not enter consciousness. The special sense organs – eye, ear, nose, tongue and skin – provide information of a particular quality and elicit complex responses. The nature of a sensation is determined by the receptor organ, not by the stimulus. The same cause will produce different effects in different sense organs, e.g. radiation may be appreciated as light by the eye but as warmth by the skin, a tuning-fork is only a touch to the skin but conveys a note to the ear. Conversely, a sense organ can only react in one way, whatever the stimulus; the eye not only reacts to light normally but also transmits the sensation of light when it is pressed vigorously.

Another characteristic of the relation between brain, sense organs and the outer world is that of *projection*. Although sensations are only felt as the result of changes in the brain cells, they are experienced as if taking place more remotely; touch is projected to the skin and taste to the mouth, while we regard sight and sound as coming from the surroundings. The brain provides a faithful *simulacrum* of the environment.

Finally, sense organs obey physical laws. There is a minimum threshold value for the stimulus, below which it is ineffective, and this level depends on physiological factors; it is higher in fatigue or if there are distracting stimuli. Further, we do not experience sensations in an intensity directly proportional to the physical intensity of the stimulus; there is such an enormous range of impressions from the outside world that we have to summarize them. A light that we feel to be twice as bright as another is by physical measurement perhaps 10 times as intense, and this applies also to sound, i.e. it is logarithmic.

The important factor in appreciating a stimulus is the change from a steady state. A fresh or intensified stimulus causes a burst of impulses which die away even though the stimulus persists – the phenomenon of *adaptation*. This is essential or we could not tolerate the continued stimulus of sunlight or the pressure of our clothes.

THE SKIN AS SENSE ORGAN

The skin intervenes between the body and the outside world, so its sensory reactions are of prime importance. The eye, ear and central nervous system itself have all developed from an infolding of this surface layer in evolution, recapitulated in the human embryo. The modalities involved include touch, pressure, pain, temperature, vibration and stereognosis.

Touch is transmitted by small end-bulbs of the cutaneous nerves, in little bays on the deep surface of the epidermis (Figure 20.14). The hairs play an important part by magnifying the effects of contact and the sensory bulbs are grouped around their roots; hence the phenomenon of *tickle*, and the blunting of sensation after shaving a hairy surface. The skin is not uniformly sensitive; some points are more so than others. There are a number of scattered touch spots separated by 0.5 mm or so, between which there is no sensation. The capacity of discriminating between adjacent stimuli varies from place to place. On the tongue and fingers, where touch spots are closely packed, two compass points are felt as separate if only 1 mm apart, but separation on the thigh must be 6 cm (2/3inch) before being felt as distinct.

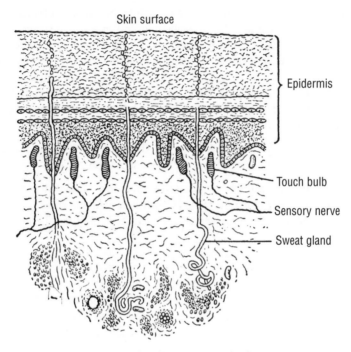

Figure 20.14 Section of the skin showing touch bulbs and sweat glands

The skin illustrates how our conventional reading of sensation depends on *habituation*. A pencil between the tips of middle and index fingers is felt to be single because these surfaces are normally adjacent and we learn to fuse their sensations; if the fingers are crossed, surfaces never normally in contact are brought together and the pencil is now felt as double.

It is doubtful if there are any really specific *pain* receptors. Pain and touch are intermingled; in the cornea of the eye any touch is felt as pain, and there is no ordinary touch sensation. The object of pain is to protect by evoking an automatic withdrawal reflex. When pain is lost in nervous diseases, a patient may burn him- or herself unawares, and the skin becomes liable to infection and ulceration from unappreciated pressure.

There is a close link between *temperature* and pain; both are transmitted along the same fibre tracts in the central nervous system. The headquarters for pain appears to be the thalamus, and 'pain' is the result of a varying analysis of stimuli, for the same sensation is sometimes painful and sometimes not, and pain is always affected by attention and emotional state. The severance of the fibre tracts connecting the frontal cortex and thalamus may render even severe pain tolerable, because it is not magnified by 'affect'.

The viscera are normally insensitive to cutting or burning, and the brain itself appears to be entirely insensitive. Then there is the phenomenon of *referred pain*, where pain originating from a deeper structure is felt over the skin surface. Thus pain from the diaphragm is often felt at the shoulder because the phrenic nerve to the muscle originates from the same segments of the cord that supply the skin of the shoulder; and a prolapsed intervertebral disc in the lumbar region irritates a root of the sciatic nerve and causes pain in the lower limb.

▶ Temperature sensation

Feelings of heat and cold are due to loss or gain of heat from the object felt. Temperature sensation is not uniformly distributed but in discrete points at the skin surface, and there seem to be separate sets of receptors – heat and cold spots. The mucous membranes are less sensitive to heat, so that one can drink fluids that are too hot to hold. Obviously, there is a qualitative difference between the hot and the merely warm, but the problem of temperature sensation is complicated by the changes produced in the small blood vessels of the skin. If these dilate, the part becomes warm because flushed with blood; constriction results in coldness. A painful stimulus often causes vasodilatation and is spoken of as a *burning* pain.

The skin is also sensitive to *vibration*, more so if the vibration is transmitted to underlying bone. Vibration is closely linked to touch and pressure sensation.

Finally, there is *stereognosis*, the ability to recognize three-dimensional shapes. For this the object has to be handled, not just felt, the fingers must be warm and sensitive, and the cerebral cortex must be able to analyse the information supplied. Naturally, it is better developed in the blind.

THE EYE

The globe of the eye lies embedded in the orbital fat, embraced by six small *ocular muscles* that spring from the walls of the orbit and move the eyeball in all directions. It is a sphere almost 2.5 cm (1 inch) in all directions, with a more local bulge at the anterior pole formed by the window of the *cornea*. The optic nerve is attached behind as a stalk that runs back into the cranial cavity via the optic foramen.

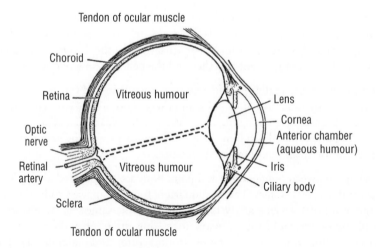

Figure 20.15 Section through the globe of the eye

The eyeball has three coats.

1   An outer fibrous coat, the tough opaque *sclera*, the white of the eye which surrounds the globe except where it merges into the cornea.

2   An intermediate pigmented layer, the *choroid*, which lines the sclera and is modified in front to form the diaphragm of the *iris*. This rests on the lens and has the central aperture of the *pupil*, the size of which is varied by contraction of the underlying muscular ciliary body.

3   The innermost, nervous, layer is the *retina*, adapted for the reception of light stimuli and extending forward no further than the ciliary body. It is pigmented, containing the *visual purple*, which is bleached by exposure to strong light, and consists of a layered pattern of nerve cells whose fibres leave the eye in the optic nerve. The point of attachment of the latter to the retina is the *optic disc* or blind spot; here the central retinal artery and vein enter the eye by traversing the optic nerve.

There are three refracting media of the eye.

1   The watery *aqueous humour* fills the anterior chamber, the space between cornea and lens, and also the small posterior chamber, a narrow cleft between lens and back of iris. Like the cerebrospinal fluid, the aqueous humour is constantly secreted and reabsorbed.

2   The translucent *lens* is a solid structure enclosed in a capsule, more convex on its posterior aspect. This curvature is modified by changes in the tension of its suspensory ligament to accommodate for near and far vision.

3   The bulk of the eye is filled with the thin translucent jelly of the *vitreous humour,* traversed by a narrow central canal running from the entry of the optic nerve to the back of the lens.

▶ **Eyelids and lacrimal apparatus**

The supporting substance of each lid is the tarsal plate, a layer of dense connective tissue. The eyelashes are attached at the lid margins and the inner surfaces are lined by a delicate membrane, the *conjunctiva*, which is reflected over the cornea and sclera. These are kept moist by the secretion of tears by the *lacrimal gland*, situated in the upper and outer part of the orbital cavity and with ducts opening under the upper lid. The secretion is collected in the *lacrimal sac* at the inner angle of the lids. From here the *nasolacrimal duct* runs down the side-wall of the nasal cavity to discharge into the nose.

▶ **The eye as an optical system**

The eye may be compared to a camera, to the former's advantage. Focusing is automatic and the lens changes its shape, or *accommodates,* to become more curved and powerful for near vision and less so for distant objects, the aim being to converge light rays exactly on the retina. A normal eye produces a sharp retinal image of an object at infinity without any accommodation at all. There is a very wide field of vision, over 200° of a circle, so that objects slightly behind eye level are still in sight, and there is little loss by reflection at the refracting surface.

The cup-shaped retina is in just the position for the best definition of the image, and it has two different systems of receptors: rods and cones. The rods, with their visual purple, are for blacks, whites and greys under twilight conditions. The cones have other pigments for colour vision in bright light; when this system is defective, the individual is colour-blind. In darkness the retina becomes more sensitive to the available light, the phenomenon of *adaptation*.

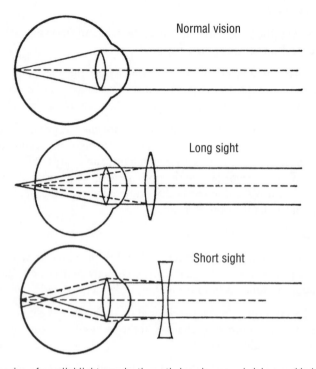

Figure 20.16 The focusing of parallel light rays by the optic lens in normal vision, and in long and short sight

When there is a refractive error an artificial lens has to be used to help the rays converge on the retina; the dotted lines show how this is achieved

Refractive errors

Longsightedness is normal in early childhood, for the eyeball is too small for proper convergence. If this persists, as *presbyopia*, in adults, light rays converge behind the retina and correction is by a convex spectacle lens to supplement the converging power of the natural lens. The long sight of old age is due to loss of resilience of the lens so that it cannot assume a more convex shape.

In shortsightedness, or *myopia*, rays are focused in front of the retina by over-powerful refraction. This is corrected by a supplementary concave lens to displace the point of convergence backwards.

Astigmatism results from the lens/cornea system being more curved in one plane than another, i.e. not truly spherical. Accommodation is correct for a given line, but a line at right angles is not properly focused. Correction is with a cylindrical lens.

Visual acuity is not a measure of the *size* of an object, for anything is visible if it emits enough light. It is measured by estimating the minimum angle subtending two lines that can be appreciated, normally this is one minute of arc.

Binocular vision is important because the combined use of both eyes gives a wide field of vision and eliminates the screening effects of nose and brow. It also provides the stereoscopic effect, which depends on the fact that the images of an object formed by each eye are slightly different, but are presented simultaneously to the brain without appearing double. To secure this, both eyes must converge on the object. In people with poor converging powers the single image is easily dissociated into two; and if there is an habitual preference for the better eye, the image of the weaker one is permanently suppressed and the eye may become virtually blind.

THE EAR

The ear has three parts: the *external ear*, the *middle ear* or *tympanic cavity*, and the *internal ear* or *labyrinth*.

The external ear is mostly outside the skull, while the middle and inner ear are within the temporal bone.

The *external ear* consists of the funnel-shaped organ for collection of sound waves, a thin plate of elastic fibrocartilage covered with skin. This leads along a narrow channel, the *external auditory meatus*, 2.5 cm (2/3 inch) long, cartilaginous at the surface but bony as it approaches the skull, and lined by skin with wax-secreting glands. At the bottom of the external meatus, the circular *tympanic membrane* – eardrum – shuts off the middle ear cavity.

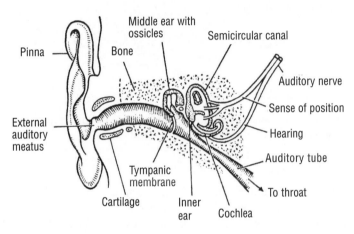

Figure 20.17 The ear

The *middle ear* is a roughly cuboidal cavity. Its roof is the floor of the middle cranial fossa, the eardrum is its outer wall, and the bony case of the inner ear lies medially. A canal, the *auditory tube*, connects it with the nasopharynx. Its function is to equalize the air pressures on either side of the drum; if the pressure is suddenly altered, as in an aeroplane ascent, there is temporary deafness until the canal is opened by yawning or swallowing. The cavity is spanned by three tiny bones, the *auditory ossicles*, which link the tympanic membrane to the outer wall of the inner ear, transmitting the vibrations set up by sound waves in the eardrum to the inner ear by bone conduction.

The *internal ear*, deep in the temporal bone, is a complex organ, the *membranous labyrinth*, housed within a correspondingly shaped *bony labyrinth*. The upper part is concerned with sense of position and spatial orientation; this consists of three fluid-filled *semicircular canals* arranged in planes mutually at right angles. Then there is an intermediate portion, the *utricle* and *saccule*. And below this is the *cochlea*, the true organ of hearing, a spirally coiled structure responding to pitch and volume.

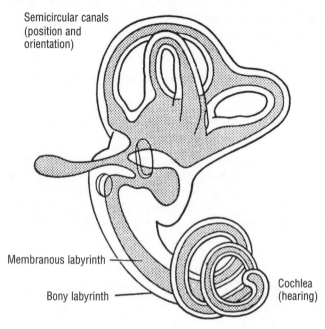

Semicircular canals
(position and
orientation)

Membranous labyrinth

Bony labyrinth

Cochlea
(hearing)

Figure 20.18 The internal ear

On the inner cranial side is an *internal auditory meatus* which allows passage of the seventh and eighth cranial nerves between the temporal bone and the brain.

▶ Gravity and position sense

The semicircular canals, utricle and saccule, are filled with fluid and lined wth hair-cells which respond to changes in position of jelly-like mineralized masses – the *otoliths* – which respond to gravitational changes due to changed positions of the head. Rotational movements stimulate the hair-cells and control of accurate eye movement to compensate for head movements depends on information supplied by the canals. Motion sickness is due to over-stimulation of the canal system. Impulses from the canals are transmitted along the vestibular portion of the eighth cranial nerve to the cerebellum, which adjusts posture in response.

▶ Hearing

The tension of the eardrum is not constant but is adjusted by a muscle which springs into contraction to protect the delicate middle ear mechanism against very loud noises. The membrane responds faithfully to high and low frequencies between 40 and 30,000 cycles per second and transmits these vibrations to the ossicles in the middle ear; these beat against the bony case of the labyrinth and stimulate the cochlea.

Hearing is effective over a wide range of pitch and intensity; it is most sensitive to frequencies around 1000 c/s. It is also direction-finding, due to the position of the ears at either side of the head, and differences in the time of arrival of sounds at either ear. Hearing impulses are transmitted along the acoustic portion of the eighth nerve.

THE NOSE

The external portion of the nose is largely fibrocartilaginous. The nasal cavity is divided into right and left halves by the septum. The cavities open externally at the nostrils or *anterior nares* and behind into the nasopharynx at the posterior nares. The side-wall of each cavity is marked by three ridges: the *superior, middle* and *inferior conchae*, the substance of which is a fragile scrolled *turbinate bone*; and beneath each turbinate is a recess or *meatus*. The accessory nasal air sinuses open into one or other meatus, the maxillary antrum into the middle meatus and the frontal sinus by a more tortuous channel. The nasal mucous membrane is continuous with that of the sinuses. The smell-sensitive region is the roof of each cavity, where the mucous membrane contains the olfactory cells whose fibres make up the first cranial nerve.

The sense of smell, though vestigial in humans, remains one of the most delicate of the senses. It depends on odoriferous particles dissolving in the moisture of the mucous membrane. The sensory area is a patch of membrane at the summit of the nasal cavities, so there is a little delay before a new odour is appreciated. The nature of the sensation aroused probably depends on the molecular configuration of the substance. Pungent 'smells' such as that of ammonia are really stimuli of ordinary sensation and are transmitted by a different nerve.

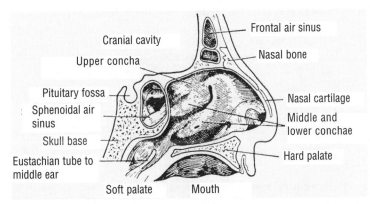

Figure 20.19 The nose, with the septum removed, and showing the side-wall of the left nasal cavity (after *Gray*)

TASTE

Taste, like smell, is a chemical sense and depends on the entry of foreign particles into solution on the tongue and palate. Microscopic *taste buds* in the mucous membrane serve this function and in the tongue these are grouped into obvious projections, or *papillae*. Taste buds are also found on the soft palate and epiglottis. Their impulses are collected up in the seventh and ninth cranial nerves. There are five basic tastes: sweet, sour, salt, bitter and the recently confirmed umami; these do not blend but remain distinguishable. The nature of any taste depends not only on the chemical composition of a substance but also on its acidity or alkalinity. Many so-called tastes and flavours are really smells, and are lost when the latter sense is put out of action during a cold.

Test yourself

1 Which part of a nerve cell interlocks with adjacent nerve cells?
   a Axon
   b Dendrite
   c Nissl body
   d Oligodendrocyte

2 Identify how a myelinated nerve fibre transmits conduct impulses.
   a Slowly
   b Rapidly
   c At moderate speed
   d It does not conduct impulses

3 What name is given to a period of time when a nerve will not transmit a stimulus?
   a Action potential
   b Electrochemical gradient
   c Refractory period
   d Reflex arc

4 Identify how many pairs of cranial nerves arise from the brain.
   a 8
   b 10
   c 12
   d 14

5 What is the outermost layer of the enveloping membrane of the brain and spinal cord called?
   a Pia mater
   b Arachnoid layer
   c Cerebrospinal fluid
   d Dura mater

6 Which area of the brain is related to the emotions?
   a Thalamus
   b Occipital lobe
   c Parietal lobe
   d Cerebral cortex

7 Which area of the brain contains the nerve centre for controlling respiration?
   a The cerebellum
   b The pons
   c The medulla oblongata
   d The pituitary body

8 Which fibre tract in the brain is concerned with voluntary motion?
   a Cerebellospinal
   b Pyramidal
   c Posterior bundles
   d Spinocerebellar bundle

9 Which area of the brain secretes cerebrospinal fluid?
  a Choroid plexus
  b Aqueduct
  c Lateral ventricle
  d Cerebellum

10 What term is used for the ability to recognize three-dimensional shapes?
  a Vibration
  b Simulacrum
  c Adaption
  d Stereognosis

21

The endocrine system

In this chapter you will learn about

▶ *different types of hormones*

▶ *the pituitary gland and its hormones*

▶ *the thyroid gland*

▶ *the parathyroid glands*

▶ *the suprarenal glands*

▶ *the pancreas and insulin*

▶ *the gonads.*

Animals have two ways of responding to stimulation. The most primitive is by an alteration of the chemical substance of the protoplasm at the stimulated point with diffusion of the chemical agent to other parts of the body. The more sophisticated method is the development of a nervous system. However, the primitive mechanism has been retained, and even developed, in the higher animals. The *endocrine system* is a group of glands whose secretions are *internal*, i.e. absorbed into the bloodstream and diffused throughout the body, on whose functions they exert a profound control. These secretions are the *hormones*, or chemical messengers, and these glands differ entirely from those of *external* secretion such as the liver and salivary glands, which have ducts opening into a body cavity or on the skin surface. Some organs are capable of both internal and external function: the pancreas secretes digestive juices into the bowel as well as insulin into the blood, and the ovaries and testicles form ova and spermatozoa as well as the male and female hormones.

A characteristic of hormones is the potency of even minute amounts. There are three main types of endocrine activity:

1 a *temporary response* to emergency, e.g. the liberation of adrenaline from the suprarenal glands

2 a *sustained activity*, e.g. the constant stimulation of metabolism by the thyroid gland

3 the secretion of the *sex hormones* by the sex organs.

The endocrine glands are largely, but not entirely, independent of nervous control, which is most important in relation to the adrenal medulla and the posterior pituitary. They influence each other's activities through the action of their hormones. *Chemically*, hormones are either:

▶ *proteins* or polypeptides, such as the secretion of the anterior lobe of the pituitary

▶ *steroids*, such as those of the adrenal cortex

▶ *phenol* derivatives, such as thyroxine and adrenaline.

Hormones maintain internal homeostasis by controlling water and electrolyte balance (adrenal cortex), blood sugar (insulin) and metabolic rate (thyroxin). They also influence metabolism and reproduction, growth and development, and responses to stress.

They have 'target' organs or tissues. These may be specific, e.g. the thyroid-stimulating hormone of the pituitary affects only the thyroid; or they may, like the adrenal corticosteroids, affect the metabolism of most cells. They activate or deactivate enzymes, excite secretory activity and stimulate mitosis.

All the endocrine glands are closely interrelated, the precise balance in the individual affects personality. They are an orchestra of which the leader is the pituitary gland.

Pituitary

The pea-sized pituitary gland is slung under the brain immediately behind the optic chiasma, occupying a little pocket in the base of the skull between the two middle cranial fossae. It is attached by a short stalk to an important region of the brain called the *hypothalamus* in the floor of the third ventricle. The hypothalamus and pituitary are in close circulatory communication. Although not essential to life, disease or removal of the gland results in arrest of growth, atrophy of the sexual organs, weakness and senility. It consists of an anterior and posterior lobe, each of which produces several hormones.

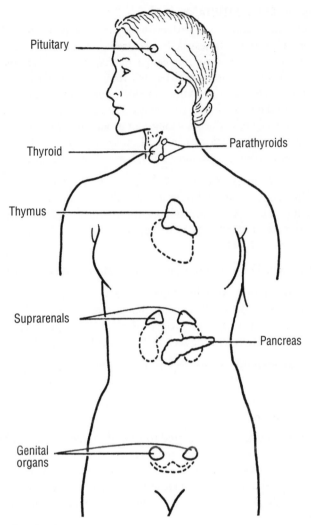

Figure 21.1 The endocrine glands

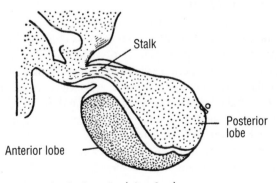

Figure 21.2 The pituitary gland, longitudinal section (after *Gray*)

HORMONES OF THE ANTERIOR PITUITARY

1 *Growth hormone.* This is concerned with growth, particularly of the skeleton. It is necessary for normal protein synthesis and increase in size of the tissues. Essentially, it promotes anabolism. It is secreted in irregular bursts, more at night than by day. Overproduction of this hormone, as may occur with a tumour of the secreting cells, causes *gigantism* in children due to overgrowth of the long bones. In adults, after epiphyseal closure, the corresponding condition is *acromegaly*, an overgrowth of the soft tissues of the hands, feet, face and tongue, with coarsening and broadening of the features. The tumour responsible may press on the optic chasma and cause blindness. Deficiency of growth hormone in childhood produces *pituitary dwarfs*, sexually immature individuals.

2 *Adrenocorticotrophic hormone (ACTH).* This is necessary for activity of the cortex of the adrenal gland. No significant steroid output is possible without pituitary stimulation, and the close connection of the two glands is sometimes termed the 'pituitary-adrenal axis'. If the pituitary is destroyed the patient can be kept in health only by administration of the adrenal hormones or of ACTH.

3 *Thyrotrophic hormone.* This is responsible for normal function of the thyroid gland.

4 *Gonadotrophic hormones.* These control the release of ova and female sex hormones by the ovary and are responsible for the cyclical changes in that organ through a follicle-stimulating hormone (FSH) and a luteinizing hormone (LH) (see Chapter 22). In the male they control sperm formation and the secretion of male hormones by the testes.

5 *Prolactin.* This is essential for the development of the breasts during pregnancy and the secretion of milk.

▶ Pituitary control system

The anterior lobe of the pituitary is stimulated to produce these hormones by releasing factors passed down from the hypothalamus, and the latter is under the control of higher cerebral centres in response to various stresses. After the appropriate pituitary hormone has reached the target organ – say, the adrenal cortex – the latter is stimulated to secrete its own hormone – in this case, cortisol – into the bloodstream. An ingenious feedback mechanism now comes into play. Once the blood level of cortisol has reached its optimum, this inhibits further production of ACTH by the pituitary. Then, as the concentration of target gland hormone falls, the pituitary is reactivated and produces more of its trophic hormone. This mechanism applies to all the secretions of the anterior pituitary and is represented as follows.

HORMONES OF THE POSTERIOR PITUITARY

1 *Antidiuretic hormone (ADH).* This controls water secretion by the kidney (see Chapter 19). It increases the permeability to water of the terminal ducts of the renal tubules and so promotes reabsorption of water into the circulation and concentrates the urine. Again a feedback mechanism comes into play, designed to maintain constant osmolarity in the blood and extracellular fluid. Should the osmotic pressure of the plasma rise, this stimulates specific *osmoreceptors* in the brain, more ADH is liberated, water is retained in the blood to exert a diluting effect and a smaller volume of urine is passed. Should the osmotic pressure of the plasma fall, ADH secretion is inhibited, a larger volume of urine is passed, and the plasma becomes more concentrated. ADH deficiency causes *diabetes insipidus*, with the output of huge amounts of urine and intense thirst.

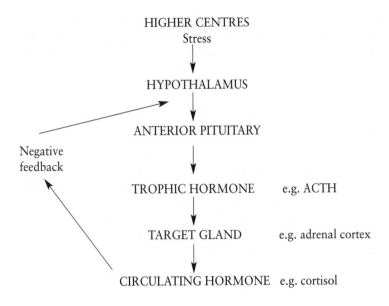

HIGHER CENTRES
Stress

↓

HYPOTHALAMUS

↓

ANTERIOR PITUITARY

↓

TROPHIC HORMONE   e.g. ACTH

↓

TARGET GLAND   e.g. adrenal cortex

↓

CIRCULATING HORMONE e.g. cortisol

Negative feedback

2 *Oxytocin.* This exerts a stimulating action on smooth muscle, notably of the uterus and the milk ducts in the breast. Oxytocin plays an important part in uterine contraction during childbirth and in ejection of milk from the breast.

Thyroid

The thyroid has two main secretions.

1 *Thyroxin.* This is secreted in response to the stimulus of the thyrotrophic hormone of the pituitary. It is a complex glycoprotein containing iodine. Dietary iodine is derived from the soil and where it is deficient the thyroid hypertrophies to try to compensate and forms a swelling, or *goitre*, in the neck. The condition is prevented by iodizing table salt.

Thyroxin stimulates metabolism; it increases the rate of tissue build-up and breakdown and consumption of oxygen and is essential to normal development. Excess of secretion causes *thyrotoxicosis* or Graves's disease – overexcitability, rapid pulse, bulging eyes and wasting. Deficient secretion in children causes *cretinism* – dwarfism and mental retardation. In adults, a deficiency causes *myxoedema* – obesity and mental disability often mistaken for dementia but eminently treatable with thyroxine.

2 *Calcitonin.* This transfers calcium from plasma to bone and densifies bone. It is used in the treatment of conditions characterized by rarefaction of bone.

PARATHYROIDS

The parathyroids are four small pea-like glands embedded in the thyroid capsule, two on either side. Their hormone acts together with calcitonin to regulate the level of plasma calcium at 9–11 mg/100 ml. Parathormone mobilizes calcium from bone into the blood and promotes intestinal absorption of calcium. It is secreted whenever plasma calcium tends to fall. Conversely, if plasma calcium rises, calcitonin comes into play. If there is an excess of parathormone, as with a tumour of the gland, so much calcium is moved out of the bones that they appear cystic and transparent in the X-ray and fracture easily. If there is not enough parathormone, as may happen when the glands are inadvertently removed during a thyroidectomy, the plasma calcium falls to low levels, causing tetanic convulsions.

Suprarenal glands

The suprarenal is a triangular cap-like organ sitting on the upper pole of the kidney. It has an outer rind, or *cortex*, and an inner *medulla*.

The function of the *medulla* is intimately connected with the sympathetic system. Its hormones – *adrenaline* and *noradrenaline* – are secreted in response to sympathetic stimuli in emergency. Their release excites the usual adrenergic effects at the termini of the sympathetic system. There is increase in the rate and force of the heartbeat, the blood pressure rises, the smooth muscle of bowel and bladder relaxes, the pupils dilate and the hair stands on end. Blood is also diverted to skeletal muscle and glycogen is mobilized from the liver to provide energy to prepare the body for fight or flight.

The internal secretions of the cortex are known as corticosteroids and are based on the carbon ring structure:

They include *mineralocorticoids*, regulating electrolyte balance, and *glucocorticoids*, influencing cellular glucose metabolism. The most important are cortisol (hydrocortisone) and aldosterone, which profoundly influence the metabolism of carbohydrates and electrolytes respectively. *Cortisol* promotes glucose formation and retention, favours protein breakdown and inhibits tissue repair. *Aldosterone* modifies the transport of sodium and potassium ions across cell membranes, favouring sodium retention and potassium excretion in the renal tubules.

Cortisol formation is increased in stress situations, not just emergencies but also those of a more prolonged nature after operations and injuries, in convalescence and emotional crises. It is produced in response to stimulation of the adrenal cortex via the hypothalamic-pituitary-adrenal axis, and if the cortex is unable to respond the individual may collapse and die.

The suprarenal cortex also has a close connection with the sex glands. Abnormal function of the cortex can lead to children of indeterminate sex at birth, or bring on precocious sexual maturation, or actually tend to *reverse* the secondary sexual characteristics, producing masculinity in women and feminism in men.

Pancreas and insulin

Insulin is the internal secretion of groups of cells in the pancreas known as the *islets of Langerhans*. It is an anabolic hormone that promotes the storage of carbohydrate, protein and fat. Insulin is secreted in response to the rise in blood sugar after a meal and acts to lower this level by facilitating the take-up of glucose in the tissues and inhibiting the formation of glucose in the liver. It also promotes glycogen deposition in muscle. Insulin deficiency causes diabetes mellitus with high blood sugar and sugar in the urine. Severe diabetes requires insulin injection, but maturity onset diabetes may be controlled by diet and oral drugs. *Hypoglycaemia* is a low blood sugar usually caused by insulin overdose and causes anxiety, weakness, possibly convulsions and death.

The *thymus* may be regarded as an endocrine gland, inasmuch as it secretes a factor essential for the development of lymphoid tissue in the lymph nodes and spleen in the newborn and the maintenance of these tissues, and the production of new immunologically competent lymphocytes in the adult.

Gonads

The ovaries and testes, besides forming the reproductive cells – ova and spermatozoa – secrete the female and male sex hormones, *oestrogens* and *androgens*. These are steroid hormones, formed in response to gonadotrophic drive from the anterior pituitary, with the usual feedback control.

These hormones are responsible for the development of the *secondary sexual characteristics* – the attributes, though not the essentials – of sexuality. These include hairiness, a deep voice, enlargement of the external genitalia, and a narrow pelvis in men. In women, these hormones develop hairlessness, smooth skin with much subcutaneous fat, breast development and a broad pelvis.

In more detail (see also Chapter 22 on reproduction), the ovarian hormones consist of two groups: oestrogens and progestagens. *Oestrogens* initiate the cyclic activity of the genital tract and breast, promoting thickening and secretion in their lining membranes and muscular development in the walls of uterus and vagina. They are more concerned with the first, preparatory or *follicular*, stage of the menstrual cycle associated with ovulation. The *progestagens* have to do with the second, receptive or *luteal* half of the cycle, in readiness for possible pregnancy, and their activity persists if pregnancy ensues.

Oestrogen also has some general metabolic effects. Its production falls off sharply at the menopause, often resulting in some emotional instability and masculinizaton, both of which can be relieved by giving the natural or synthetic hormone.

The most important androgenic hormone is *testosterone*. This stimulates the growth and function of the male reproductive tract, including normal sperm formation. It also has a marked anabolic effect on general metabolism, promoting protein synthesis, growth, and muscle and skeletal development, which is why it is sometimes abused by athletes of both sexes.

Finally, and paradoxically, ovary and testis also secrete small amounts of the *opposite* sex hormones.

Test yourself

1   Identify a hormone produced by the anterior pituitary gland.
    a   Insulin
    b   Gonadotrophic hormone
    c   Antidiuretic hormone
    d   Oxytocin

2   What is the role of thyroxine?
    a   It exerts a stimulating action on smooth muscle
    b   It controls water secretion by the kidneys
    c   It is concerned with the growth of the skeleton
    d   It stimulates the metabolism

3   Which mineral is required for the production of thyroxin?
    a   Calcium
    b   Magnesium
    c   Iodine
    d   Zinc

4   Identify a hormone that helps to transfer calcium into bone.
    a   Insulin
    b   Prolactin
    c   Parathormone
    d   Calcitonin

5   Which of the following is a response to the release of adrenaline?
    a   Decrease in blood pressure
    b   Increase in heart rate
    c   Contraction of the smooth muscles in the bowel
    d   Constriction of the pupils

6   What is the function of cortisol?
    a   It promotes glucose breakdown
    b   It promotes protein breakdown
    c   It promotes fat breakdown
    d   It promotes glycogen breakdown

7   Where in the body is cortisol produced?
    a   Cortex
    b   Medulla
    c   Pituitary
    d   Thyroid

8   What is the main method by which hormones are transported to the organs upon which they act?
    a   In the bloodstream
    b   By muscle contraction
    c   By the nervous system
    d   By the digestive system

9 Where is the thyroid?
   a Brain
   b Kidneys
   c Neck
   d Groin

10 Which hormone influences the metabolism of electrolytes?
   a Cortisol
   b Aldosterone
   c ADH
   d Insulin

Reproduction and development

In this chapter you will learn about:

▶ *the male and female reproductive organs*

▶ *reproduction*

▶ *the development of the embryo*

▶ *breasts and milk production*

▶ *genes and inheritance*

▶ *genes and the environment*

▶ *gene mutations.*

Reproductive organs

The essential sex organs are the *gonads*, which are a pair of *testes* in the male forming the spermatozoa, and a pair of *ovaries* in the female forming the ova. The general plan of the reproductive system in both sexes is not dissimilar, but the ovaries remain in the abdominal cavity while the testes lie outside it. And in the female there is a uterus to house the developing embryo, while the penis of the male is represented in the female by the diminutive clitoris.

MALE ORGANS

The *testes* develop in the abdominal cavity but pass down before or just after birth to enter the loose skin pocket of the scrotum externally, where they hang down on each side of the root of the penis. There remains an oblique passage through the abdominal muscles just above the inguinal ligament, the *inguinal canal*, occupied by the stalk of the testis or *spermatic cord*, which carries blood vessels to the organ. The *ductus deferens* carries blood back to the pelvis. This passage is a potentially weak point for the development of hernia.

The testes are ovoid, with a tough capsule made up of lobules containing the fine *seminiferous tubules* in which the spermatozoa are formed. Applied to their outer side is a curved organ, the *epididymis*, which receives sperm from the testis. It is an intricately coiled tube from which issues the ductus deferens, which is the main spermatic channel and runs up to the abdomen. Testes and epididymis lie vertically in the scrotum, surrounded by a loose serous sac.

The ductus deferens runs down the side-wall of the pelvis to reach the back of the lower part of the bladder, where it lies on the upper surface of the prostate. Here a sac is attached to its outer side, the *seminal vesicle* for the storage of sperm (Figure 22.3). The ducts of the testes and vesicle open via a common *ejaculatory duct* into the urethra, which lies embedded in the prostate gland. Sperm are continuously formed in the testes, stored in the vesicle, and only enter the urethra in the ejaculation of orgasm, when they are discharged via the urethra and out of the penis.

The ejaculated fluid is not just the sperm but a complex fluid also containing the secretions of vesicles and prostate. The *prostate* is a solid organ of muscular and glandular tissue the shape and size of a chestnut. It has its base applied to the neck of the bladder and its apex on the pelvic floor, and is traversed by the urethra.

The *penis* consists of a central bulb arising from the centre of the perineum, traversed by the urethra, and two lateral *crura* springing from the sides of the bony pubic arch. These join to form the shaft of the organ, a cross-section of which is shown in Figure 22.4. Here, the part containing the urethra, the continuation of the bulb, is the *corpus spongiosum* below, with the *corpora cavernosa*, continuations of the crura above on each side. It is to these latter that the organ owes its property of increase in length and girth on sexual excitement, becoming rigid. This process of *erection* is due to a system of cavernous spaces which can be rapidly distended with blood from the penile arteries.

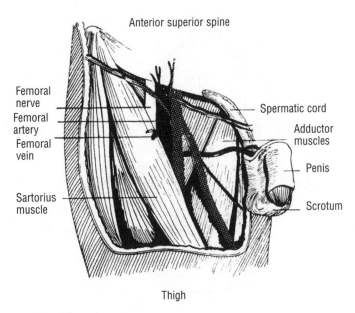

Figure 22.1 Contents of the right groin

Note the femoral nerve and vessels, the associated muscles and the spermatic cord connecting the testicle in the scrotum with the abdomen (after *Gray*)

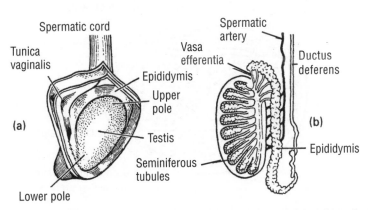

Figure 22.2 (a) Testis and epididymis exposed by reflection of their serous covering, the tunica vaginalis; (b) the same in longitudinal section

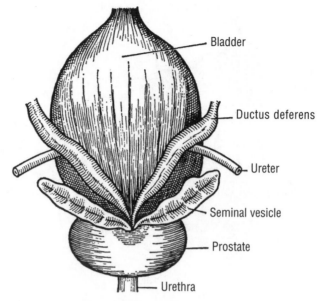

Figure 22.3 Bladder and associated organs in the male (posterior aspect)

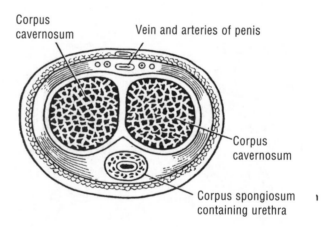

Figure 22.4 Cross-section of shaft of penis (after *Gray*)

FEMALE ORGANS

The *ovaries* are a pair of almond-shaped organs lying on the side-wall of the pelvis just below its brim. They are studded with the ovarian *follicles*, in which the egg cells or ova ripen, one coming to maturity each month. All the ova that will ever exist are already present at birth, although in a state of immaturity, whereas spermatozoa are continually formed throughout a man's life. The ovaries are embedded in the *broad ligament*, which stretches from the uterus to either side of the pelvis. In the upper free edge of this ligament are the *uterine tubes*, or *fallopian tubes*, which are attached to the uterus like outstretched arms. These muscular channels open into the uterus medially and have at their outer ends a fringed, funnel-like entrance, which embraces the ovaries so as to receive the ova when shed.

The *uterus*, or *womb*, is a hollow organ with thick muscular walls, lying in the pelvis between the bladder in front and the rectum behind (Figure 8.25). It communicates below with the vagina and at each side with the uterine tubes. It is tilted forwards so that its anterior surface rests on the bladder. Both surfaces and the dome or *fundus* are covered with peritoneum, and the peritoneal pouch between uterus and rectum is the deepest part of the abdominal cavity.

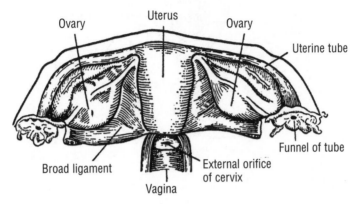

Figure 22.5 Internal sex organs of the female (after *Gray*)

The virgin uterus is on average 7.5 cm (3 inches) long, its wall 2.5 cm (1 inch) thick. Longitudinal section shows the upper *body*, 5 cm (2 inches) long, with a triangular cavity, and the lower *cervix*, 2.5 cm (1 inch) in length and traversed by a narrow canal opening into the main cavity at its internal orifice and into the vagina at its external orifice. The lining mucous membrane of the body, or *endometrium*, contains mucous glands and undergoes striking changes during the menstrual cycle. The organ becomes enormously enlarged in pregnancy but returns almost to its previous size after childbirth. The uterus is small and undeveloped in childhood and atrophies in old age.

The *vagina* is a distensible canal, capable of receiving the penis during intercourse and of allowing passage of the child in parturition. It extends from the uterus, which it meets at an angle of 90°, to run down and forward through the pelvic floor and open externally on the perineum. Part of the cervix protrudes into the vaginal vault (Figure 22.5), the encircling rim of which is known as the *fornix*. The front wall of the vagina is blended with the back of the bladder and urethra. The back wall of the vagina is separated from the rectum by fibrous tissue.

The *external female genitalia* are known as the *vulva*. This includes the *mons pubis*, a fatty hair-covered eminence in front of the pubic symphysis. The skin folds forming the lips of the vaginal orifice comprising the outer thick *labia majora* and the inner slender *labia minora*. The *clitoris* is a diminutive but sensitive erectile equivalent of the penis, and lies at the meeting of the labia in front. The *hymen* is an incomplete partition of mucous membrane stretching across the vaginal orifice in the virgin, which is ruptured by intercourse. The external opening of the *urethra* is just behind the clitoris and immediately in front of the vagina.

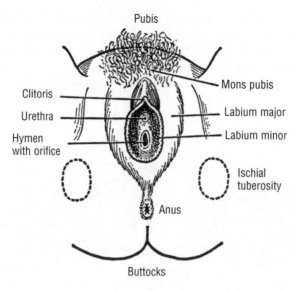

Figure 22.6 The female perineum

Reproduction

In most mammals, sexual desire usually alternates with seasons of indifference. Even when the male is continuously active, the female is receptive only in cycles which are dependent on an intermittent rhythm of ovarian activity. In humans, sexual pressure is pretty constant, particularly in the male, as is secretion of sex hormones. There is still some periodicity in the female, associated with monthly *menstruation*. Menstruation is a flow of blood and mucus from the womb.

Development of the sexual organs before puberty and their functioning thereafter depend on the pituitary gland. But when puberty arrives it is the hormones of the sex organs themselves that cause the development of the secondary sexual characteristics and the origin of sexual desire: in girls the swelling of the breasts and the commencement of menstruation, in boys the change of voice and enlargement of the external genitalia. All these changes depend on the primary development of ovaries and testes.

THE MENSTRUAL CYCLE

This is a complex pattern governed by both pituitary and ovarian hormones. It is usually a 28-day cycle beginning with the first day of menstrual bleeding, though longer and shorter cycles are common. Failure to menstruate – *amenorrhea* – may be due to developmental disorders such as absence of a uterus, hormonal upsets, starvation, athletic activity, and of course to pregnancy.

In the 28-day cycle, *ovulation* occurs at or around the fourteenth day. An ovarian follicle ripens and discharges the ovum into the peritoneal cavity. At this time there is a rise in body temperature and the woman is more sexually receptive. The egg cell is taken up by the uterine tube and propelled by muscular peristalsis towards the uterus. Here the lining becomes engorged and thickened in preparation for the embedding of a fertilized ovum. Fertilization, if it has occurred, takes place in the uterine tube. If the ovum has not been fertilized, the mucous membrane of the uterus is shed in the bleeding of menstruation. This lasts three to five days and then a new lining is built up and is completed by the fourteenth day of the cycle, midway between two periods, just when ovulation occurs again. Thus the cycle affects ovaries and uterus and is divided into:

▶ the *follicular phase* – the first two weeks
▶ a *luteal phase* – the second two weeks.

The hormonal factors are as follows. The ripening of the ovum in its follicle and its discharge at ovulation are under the control of the anterior pituitary hormone, the gonadotrophin known as FSH (follicle-stimulating hormone). Once the egg has left the ovary, the remains of the follicle become organized into a bright yellow structure, the *corpus luteum*, under the influence of another pituitary gonadotrophin – LH, or luteinizing hormone. The follicle itself has a hormone which stimulates thickening of the endometrium prior to possible conception, and the corpus luteum has a hormone which completes these preparations. Should pregnancy occur, the corpus luteum enlarges and persists, otherwise it disintegrates. Thus ovulation and menstruation alternate at fortnightly intervals. The uterine lining is completed and receptive during the latter half of the cycle, destroyed at menstruation, and rebuilt in the first half.

If the fertilized ovum embeds and develops in the uterine tube instead of the uterus, there is an *ectopic pregnancy*, which can result in rupture and life-threatening haemorrhage after a few weeks. Ectopic pregnancy can also occur in the abdominal cavity, but this is rare.

FERTILIZATION

Fusion of the male spermatozoön with the ovum normally occurs in the uterine tube as the egg is moving towards the uterus. It can only occur if a living sperm derived from the male by recent intercourse has made its way up the vagina, through the cervix and body of the uterus into the uterine tube. Success therefore depends on a near coincidence of intercourse and ovulation, approximately within 48 hours, so that the likelihood of any single sexual act resulting in pregnancy is small. When the male ejaculates, a pool of semen is deposited in the vaginal vault around the cervix. The sperm have to penetrate the cervical mucus, and upward progress is no more than 3 mm/min. Their overall progress in the uterus and uterine tube is much faster, aided by muscular propulsion in these organs. During their ascent there is an enormous reduction in sperm count. Around 100 million spermatozoa are deposited in the vagina, but only a million enter the uterus, and approximately a hundred reach the ovum. Spermatozoa do not remain active and fertile in the female tract for more than two days. Once the ovum has been penetrated by a spermatozoön, rapid changes ensue. The nuclei of the two cells fuse and the ovum becomes impenetrable to other spermatozoa.

The sperm cell has a flattened pear-shaped *head* containing the nuclear material, connected by a short *neck* to the long *tail* whose lashing movement gives the cell its mobility. Sperm formation in the testes is a complicated process and many spermatozoa are malformed and unsuitable for fertilization. If sperm are very few or even totally absent from the seminal fluid – *azoöspermia* – this is one possible cause of sterility. Spermatogenesis is also impaired if the scrotum is kept too warm or the testes are undescended.

Female sterility may be due to the inability of the uterus to house a developing embryo or the uterine tubes may be blocked. In some cases there is failure of proper maturation of the ovum and this can be treated by administration of the pituitary follicle-stimulating hormone, though with the risk of multiple births. In either sex, however, the sex organs may be abnormal, diseased or absent. In the male, the sperm may be few or sluggish, malformed or non-viable. A female's vaginal secretion may be lethal to some sperm. Fertilization in vitro (outside the body) with implantation of the early cell mass and its many variations are possible ways to address low fertility levels in males and females; however, they will not be discussed here.

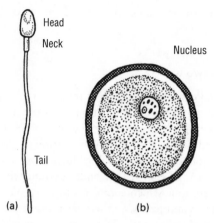

Figure 22.7 Male and female reproductive cells: (a) spermatozoön; (b) ovum

MEIOSIS

The normal process of cell division in the body cells has been described in Chapter 2 and involves splitting of each of the 46 chromosomes (23 pairs of similar chromosomes) to allocate the genetic material equally to both daughter cells. This is the *diploid* state, but if applied to ovum and spermatozoön the fertilized ovum would contain 92 chromosomes. Hence, there is a *reducing division* in germ cell formation – *meiosis* as opposed to *mitosis* – when one of each pair of chromosomes, instead of replicating by division, is simply allocated to one of the two new cells, the final germ cell containing 23 chromosomes, the *haploid* condition. Forty-four of the somatic cell chromosomes, or *autosomes*, consist of 22 matching pairs, so that their halving in the reduction division is simple and equivalent. The remaining pair, however, consists of a large X and a much smaller Y chromosome. The body cells of females have two X chromosomes and are designated XX, so the ovum formed in meiosis always has one X chromosome. Males have one X and one Y chromosome (XY) in each body cell, so half the sperm formed by the reducing division have an X and half a Y. The germ cell contains one of these but not both. If the sperm conveys an X chromosome to the ovum, the fertilized egg will contain the female XX pattern and the embryo will develop ovaries. If the sperm carries the Y, the embryo will be male (XY) and develop testes.

It is interesting that one of the two X chromosomes in the early female embryo is rendered inactive and plays no functional part for the rest of life. This may be to restore some balance between the many genes on the X chromosome compared with the few on the Y chromosome.

Chromosomal sex is the final arbiter of gender, which is not related to delayed development or abnormalities of the external or internal genitalia and certainly not to an individual's conviction that he or she really belongs to the opposite sex.

Genetically, the spermatozoön's contribution of DNA to the fertilized ovum is from the nucleus, which constitutes almost all of the cell. But the ovum contains a great deal of mitochondrial DNA in the cytoplasm in addition to the nuclear content, and this transmission of genetic information is entirely in the maternal line.

LIFE CYCLE IN MALE AND FEMALE

Adolescence denotes a state of physical and social maturity and cannot be allotted any definite period. *Puberty* means the capacity for sexual procreation – in the female that ovulation has begun and in the male that emissions of spermatozoa are occurring. The age at which regular uterine bleeding begins is called the *menarché* and occurs at an average age of thirteen, but ranging between nine and sixteen years. The onset of puberty is a complex hormonal interaction: hypothalamus, anterior pituitary ovaries and adrenal cortex all play their part.

The cessation of menstruation is the *menopause*, which occurs between the ages of 40 and 55. Many bodily and mental changes, known as the *climacteric*, are associated with this, especially osteoporosis. The ovaries cease to produce either ova or oestrogens, the genital tracts and breasts atrophy, with widespread physiological and psychological disturbance. This may be prevented or reversed by hormone replacement therapy.

In the male, sperm formation begins after the age of ten or eleven, and with this go enlargement of the penis and scrotum and the later development of pubic hair. The voice breaks and the sweat and sebaceous glands overact, so that acne is common. Skeletal growth accelerates, and this growth spurt lasts until the closure of the epiphyses of the long bones, a year or two later than in girls – hence the greater average height of men. It is difficult to pinpoint the onset of puberty in boys because there is no obvious guide such as menstruation. It is usually at between ten and fifteen years of age. The basic mechanism of hypothalamic-pituitary-adrenal-gonadal control is similar to that in the female. Once established, testicular function continues for the rest of life and is only slightly diminished in old age. There is no sharp cut-off of gonadal function as in women. Nevertheless, at about the corresponding age many men do suffer some waning of desire and physiological disturbance, suggesting that there is such a thing as a male climacteric.

Development of the embryo

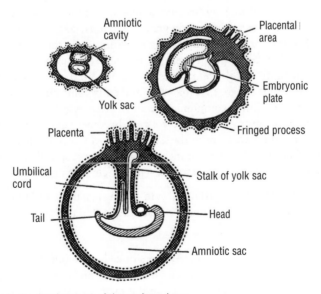

Figure 22.8 Stages in the development of the early embryo

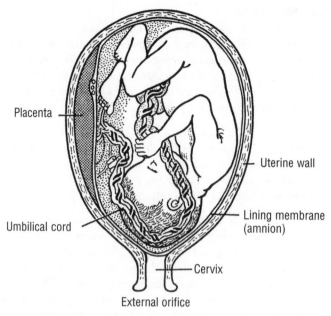

Placenta

Uterine wall

Lining membrane
(amnion)

Umbilical cord

Cervix

External orifice

Figure 22.9 The foetus in utero (after *Gray*)

The nuclei of the two germ cells fuse and the egg begins to divide to form a rounded mass of cells whose number doubles at each division. This process begins before entry ino the uterus, the journey along the uterine tube taking several days. Once in the uterine cavity the ovum adheres like a parasite (which it technically is), excavating a cavity in the mucous membrane in which it becomes embedded.

The early embryo consists of little more than a pair of sacs or vesicles, the *amniotic cavity* and the *yolk sac*, with an intervening *embryonic plate* which is the actual area at which development occurs. The amniotic cavity soon greatly expands, coming to line the entire uterus, and is filled with the amniotic fluid bathing the embryo. At the site of implantation the amniotic membrane develops a complex system of fringed processes penetrating the uterine wall. This comes to constitute the *placenta*, a fleshy disc of 15–20 cm (6–8 inches) diameter when maximally developed. It is here that maternal and embryonic circulations are in closest proximity, though there is no direct communication. Instead, oxygen and nutrients must diffuse across the separating layer of cells. The yolk sac becomes squeezed by the growth of the amnion into a narrow tube, the core of the *umbilical cord* which connects the developing child to the placenta. The cord is 51 cm (20 inches) long at birth, spirally twisted, and carries in its jelly-like substance the two umbilical arteries and single umbilical vein of the foetus.

Since the foetus sends its stale blood to the placenta for oxygenation and nutrients, the umbilical arteries contain venous blood and the vein fresh arterial blood.

EMBRYONIC LAYERS

The early embryo soon comes to possess three layers of cells. The outer layer or *ectoderm* forms the skin, its glands, hair and nails, the nervous system, and the essential parts of the eye, ear and nose. The nervous system originally lies on the surface of the back of the body as a *neural plate* and is infolded in development as the *neural tube*, the precursor of the spinal cord, with a bulbous expansion at the head end marking the brain. All the sense organs originate in this layer, by an infolding of modifed skin surface meeting an outgrowth from the central nervous system. This pattern is well shown in the eye, where the retina, an extension from the brain on the stalk of the optic nerve, cups the transparent lens developed from the overlying skin.

The innermost layer, or *endoderm*, forms the bowel and its associated glands – such as liver and pancreas – the lining of the respiratory tract and lungs, and the thyroid and parathyroid glands. The intestine develops on the posterior wall of the abdominal cavity and comes to lie suspended by a mesentery still attached to the posterior wall. The large bowel undergoes a complicated rotation to bring the caecum and ascending colon to the right and the descending colon to the left.

The intermediate layer is the *mesoderm*, which is essentially segmental, forming the vertebrae and primitive muscle blocks (*myotomes*) and *dermatomes* of the skin. It is the origin of the connective tisues of the body: bone and cartilage, the teeth, the muscles, both voluntary and involuntary, the heart and blood vessels, and the urogenital system.

At any point in the complex pattern of development things may go wrong, or not far enough. The spinal cord may be left on the surface of the back as an open layer; or it may develop normally but the two halves of the vertebral column fail to enclose it completely, leaving a gap behind, the condition of *spina bifida*. The palate may remain cleft if its two halves fail to meet in the midline; the septum of the heart may be incomplete, allowing a disabling admixture of venous with arterial blood; a bone or a limb may fail to develop in whole or part; fingers and thumbs may be too few or too many. The viscera may develop on the wrong side as a mirror image of the normal. What is remarkable is that any error is a rarity.

TIME SCALE

The two primitive vesicles show up 10 days after fertilization and the intervening embryonic plate begins to develop in the third week. The head and tail folds, neural groove and heart are obvious by the fourth week. In the fifth week appear the lens of the eye, the rudiments of the face, the gill arches and the stumps of the limb-buds and the embryo is now a 0.5 cm (1/5 inch) long. The sixth week sees the body curved on itself, the head approximating to the long tail, the umbilical cord attached to the belly nearer the latter; the liver enlarges and the limb-buds grow out and are demarcated into segments.

By the end of the eighth week the embryo is 2.5 cm (1 inch) long and is now known as the *foetus*; eyes, ears and nostrils are formed, the external genitals differentiated, and the fingers and toes marked out. Fine downy hair appears in the fourth month, when the foetus is 20 cm (8 inches) long; in the fifth month foetal movements begin and the skin becomes covered with a greasy secretion.

By the seventh month the eyelids have opened and the testicles descend into the scrotum; and though the foetus may now be viable if born prematurely, it is not adequately clothed with subcutaneous fat until the end of the ninth month.

At birth the uterus contracts, rupturing the membranes and expelling first the amniotic fluid and the foetus, then the placenta, and finally the membranes. Certain important changes occur in the child at *birth*. During foetal life, the lungs have remained unexpanded and airless; they need no blood and the stream in the pulmonary artery is shunted through a bypass directly into the aorta. With the first cry the lungs expand, the pulmonary circulation is established and the bypass closes off. The parts of the umbilical vessels contained within the foetus also become obliterated and the parasite has become an independent organism.

Breasts and milk production

The mammary glands, designed for milk secretion, form two large rounded eminences between the skin and deep fascia on the front of the chest, overlying the pectoral muscles. They extend from the second to the sixth ribs and from beside the sternum to the axilla. Small before puberty, they develop with the uterus, hypertrophy in pregnancy and especially in lactation after delivery, and atrophy in old age. There are minor cyclical changes with menstruation.

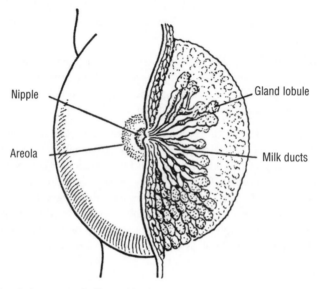

Figure 22.10 The female breast, partly dissected to show the lobules

The pigmented nipple at the apex of the breast is surrounded by a coloured area of skin, the *areola*, which is studded with sebaceous glands, is pink in the virgin, and permanently darkened in the first pregnancy. The nipple is perforated by fifteen to twenty milk ducts opening at its apex; it is sensitive and contains erectile tissue. The gland is divided into lobes by fibrous partitions, some of which fasten the organ to the chest wall. Each lobe contains a branched system of secreting gland spaces supported by fatty tissue; the larger ducts converge on the nipple and expand just before reaching it as the milk sinuses or reservoirs. The gland spaces are only distended with milk during active lactation.

The breasts develop throughout pregnancy in preparation for lactation. The first few days after delivery they yield only a watery fluid or *colostrum*, containing a few fat globules; only after this does milk secretion proper begin. This is governed by pituitary and ovarian hormones and there is a reciprocal relation between lactation and menstruation; resumption of the periods after childbirth may be delayed by continued suckling. Suckling is a powerful stimulus to the involution of the uterus.

The *milk* is a slightly acid fluid, opaque from the presence of finely dispersed fat globules. Its main constituents are:

1 *protein* (caseinogen), which is converted into a solid casein clot or curd in the stomach, and from this can be expressed a liquid whey

2 *sugar* (lactose)

3 *fat*.

In humans these are found in the proportions of: protein 1.5%, lactose 6.5% and fat 3.4%. Cow's milk contains more protein and less sugar, the fat content being similar. It can be approximated to human milk by adding water to dilute the protein content and making up the sugar, but it is a second-best substitute, and the benefits natural feeding confers on both mother and child should make breast feeding the first choice.

Genetic aspects

Chromosomes are paired and so are their genes. Because the diploid genome of the fertilized ovum contains a set of instructions from each parent, a person has two genes controlling any inherited feature. These are at the same site (*locus*) on similar chromosomes and are called *alleles*. The appearance or development of these features is known as the *expression* of the relevant gene(s). If both alleles are exactly alike, the person is *homozygous* for that feature; If they are unlike, *heterozygous*. One allele may suppress the expression of the other and is then called *dominant*, the other being *recessive*. A dominant allele will be expressed whether single or double. A recessive allele must always be double, i.e. homozygous, to manifest an effect. The *genotype* refers to whether the individual is homozygous or heterozygous for the various gene pairs. How these are actually expressed in the body is called the *phenotype*.

During maturation of the germ cells, or *gametes*, maternal and paternal chromosomes are randomly distributed at the reducing division of meiosis (see above), i.e. the members of allele pairs are segregated, distributed to different gametes. Each gamete has only one allele for a particular inherited feature, representing only one of the four possible parental alleles. Thus the number of resulting gamete types may be very large, the actual number being 2n, where n is the number of homologous pairs. Thus if the cell has eight pairs, there will be 256 possible gamete types, and in the individual testes there may be millions of possible varieties. This is a process of random assortment and is less marked – though still very considerable – in the ova, since their reducing divisions in a lifetime are much fewer. Further, genes may cross over between paternal and maternal chromosomes during the reducing division to produce *recombinant* chromosomes, thus scrambling inherited features.

In view of all these factors, the *zygote* – the cell resulting from fertilization – is only one of some 72^{12} possible variants, so that what is remarkable is the similarity rather than the dissimilarity between family members. All the genes of the zygote come from the parents, but for each child they are combined in a different way. In fact, for any particular trait, the probability of having two children of the same genotype is only around 6%.

It is this enormous randomness associated with sexual reproduction that carries evolutionary advantages by providing almost infinite variation for natural selection to act on. Each sperm and ovum and every individual (except for identical twins) is genetically unique. Chance determines the genetic make-up of the fertilized ovum.

Dominant genes may be *lethal* and cause death before or soon after birth and so are not transmitted to following generations. Other disadvantageous dominant genes, such as that for Huntington's chorea, have a delayed action; a person may survive to reproduce and his or her offspring have a 50% chance of inheriting the gene. Such genes operate for astigmatism, polydactyly, achondroplastic dwarfism and many other conditions.

Recessive genes cause other disorders such as cystic fibrosis, but will not be expressed unless they are homozygous. But parents with a single recessive allele, though not themselves affected, may transmit the gene to their children and are called carriers. This is the basis for the prohibition of marriage between close relatives, where the chance of inheriting two harmful identical genes is high. Severe genetic disorders are more common in inbred families or animals.

Dominance is not always clear-cut but may be incomplete, i.e. there is *intermediate* inheritance, the phenotype of the zygote being intermediate between those of completely dominant and completely recessive individuals. One instance is sickle-cell disease.

SEX-LINKED INHERITANCE

A trait due to genes on the sex chromosomes is said to be *sex-linked*. Unlike the other chromosome pairs, the X and Y chromosomes are not identical. The male Y chromosome has less than 20 genes, that of its X partner over 2500, so that some segments of the latter have no counterpart on the Y chromosome. A gene found only on the X chromosome is called X-linked. This means that if a man inherits an X-linked recessive gene, e.g. for haemophilia, it is always expressed because there is no corresponding dominant gene on his Y chromosome to mask it. A woman would have to have two X-linked recessive alleles to express such a disorder, so very few have X-linked conditions. X-linked traits are passed from mothers to sons, but not from fathers since males receive no X chromosome from their fathers. If one allele of a gene on the X chromosome is abnormal, it will be expressed in every male who receives it. If a woman carrying the gene marries a normal man, there is a 50% probability that her sons will show the abnormality, but none of her daughters will. Apart from single-gene inheritance, many phenotypes are produced by different genes at different sites acting together, i.e. there is *polygene inheritance*. Also, we have mentioned (see above) that some genes, transmitted by the mitochondria of the cytoplasm of the ovum outside the nucleus and its chromosomal system, are altogether and exclusively maternal. Defects in such genes are responsible for rare neuromuscular disorders. Finally, the same allele can have widely different expressions depending on whether it comes from the father or the mother, so-called *genetic imprinting*.

GENES AND ENVIRONMENT

Genes may be destiny, but they are not the whole of destiny. Environment affects the phenotype both within the uterus and after birth and through life though the genotype is unchanged. Drugs taken by the mother during pregnancy may cause gross malformations which are not genetic, as in thalidomide children. Maternal rubella may do the same. Gene expression in the individual may be modified or suppressed by environmental factors or by actions taken by the individual. A disease like spina bifida is due to a combination of genetic and nutritional factors.

Nutritional and hormonal influences after birth can cause abnormalities that nullify gene expression. If iodine is lacking in the diet, cretinous dwarfism will override a gene for height. A woman may be aware that she carries the genes for breast cancer and decide to help herself by having a mastectomy, but this will not help her female children, who will have to take their own decisions later. A man with inherited hypercholesteraemia can control his diet and take cholestatic drugs or opt for surgical procedures. In phenylketonuria, mental retardation is due to brain damage because a gene dealing with the handling of the amino acid phenylalanine is defective. Tests can be done for this at birth or before and the disease controlled by dietary methods.

Again, diseases which are undoubtedly genetically influenced, such as heart disease and breast and lung cancer, vary greatly in different countries. Breast cancer and heart disease are uncommon in Japan, but become common in Japanese who emigrate to the west and are exposed to different nutritional and other environmental factors.

In other words, though the genotype is inflexible, the phenotype is malleable. This shows the value of routine *genetic screening* for actual and potential disorders, whether obviously physical as in Down's syndrome (where there are three instead of two chromosomes 21) and congenital dislocation of the hip, or biochemical as in metabolic disorders. Such testing may be done by simple blood or saliva tests for DNA. Or the foetus can be tested by amniocentesis to examine the amniotic fluid or by sampling the chorionic villi of the placenta, though with some risk. Premarital screening is useful in detecting carriers if there are family histories of inherited diseases.

Gene therapy is emerging whereby a defective gene may be replaced by a normal gene by injecting the defective cells with viruses carrying a functional gene, or by injecting DNA directly into the patient's cells.

MUTATION

Mutations are acquired changes in the genes, and as such are the raw material of evolution because they produce 'errors', some of which may be advantageous to the organism and will be naturally selected for survival and reproduction. It is errors and their recombination or shuffling within the gene pool of the species that have made life possible. Sometimes there are known causes, e.g. ultraviolet or other forms of irradiation, free radicals, chemical toxins, cigarette smoking and other carcinogens. DNA is probably inherently at risk of errors in duplication, though there are elaborate mechanisms for their repair; however, these methods tend to fail with increasing age. The damaged segment is 'snipped out' by enzymes and the gap repaired. Efficiency in such repair is what determines length of life. Errors accumulate and if they are left unrepaired in age these may cause cancers.

Scores of enzymes take part in the formation of the simplest protein molecule, and they must be in the right place at the right time, and conform strictly to the pattern laid down in the gene. An error in the gene will cause an error in the enzyme it controls and this may initiate secondary errors.

Further, as well as these structural genes, there are *regulatory genes* which switch the former on or off, and the DNA of these genes is itself liable to error. A small informational error is usually recessive and probably has a trivial effect but, rarely, error in a single unit will change one of several hundred amino acids in a protein and cause disastrous disease (or a marked improvement). This may occur when two affected alleles of the same gene come together, one from each parent. And such errors persist and are inherited if they are in the germ cells, not in the somatic cells. Acquired mutations, as in the skin cells due to sunlight, may do minor damage in the individual, as by causing freckles, or lead to cancer, and are handed on to their daughter cells, but not to subsequent generations. Mutations in the developing spermatozoön or ovum can affect the race, for good or ill.

Test yourself

1 Which organ receives sperm from the testes?
   a Ductus deferens
   b Seminiferous tubules
   c Epididymis
   d Inguinal canal

2 What is the name of the sac that stores the sperm?
   a Spermatic cord
   b Prostate
   c Urethra
   d Seminal vesicle

3 When are the ova formed in a female?
   a In puberty
   b In the first few years of life
   c Before birth
   d After puberty

4 What is the name of the lining of the uterus?
   a Endometrium
   b Cervix
   c Fundus
   d Peritoneum

5 Which gland is responsible for the development of the sexual organs?
   a Thymus
   b Pancreas
   c Pituitary
   d Adrenal cortex

6 What name is given to the first two weeks of the menstrual cycle?
   a Follicular
   b Luteal
   c Corpus luteum
   d Ovulation

7 Identify one cause of sterility.
   a Scrotum is kept too cold
   b Scrotum is kept too warm
   c Excessive production of sperm
   d Clear uterine tubes

8 What is the reducing division in germ cell formation called?
   a Mitosis
   b Meiosis
   c Diploid state
   d Autosomes

9 What is the name for the outer layer of skin of an embryo?
 a Ectoderm
 b Endoderm
 c Dermatome
 d Mesoderm

10 Identify a sex-linked inherited condition.
 a Heart disease
 b Haemophilia
 c Down's syndrome
 d Spina bifida

23

Environmental considerations

In this chapter you will learn about:

▶ *stress under extreme conditions*

▶ *the skin*

▶ *temperature regulation*

▶ *exposure to extreme heat and cold*

▶ *the biochemistry of muscle contraction*

▶ *exposure to immersion, desert conditions, high altitudes, barotrauma and starvation*

▶ *inherent biological rhythms.*

Stress under extreme conditions

Humans are able to withstand severe climatic rigours, ranging from subzero to desert conditions, and to maintain their own body temperature and functions within fairly narrow and constant limits. This capacity to live and work under widely differing environmental conditions depends partly on clothing and cultural and social adjustments, e.g. housing, but it is mainly affected by regulating the heat produced by muscle and the heat lost from the skin.

The central organs of the body remain at a constant core temperature of 37°C, while that of the subcutaneous fat and skin surface varies widely according to their distance from the heart and the ambient temperature, ranging between 27°C and 34°C. Heat loss is mainly by radiation and convection from the skin surface. There is little loss by conduction, both skin and air being poor conductors. The air temperature alone is not an adequate measure of environmental stress as both solar and body radiation must be taken into account.

THE SKIN

The skin is protective and in mammals it also helps to regulate the body temperature since heat loss occurs almost entirely from the skin surface.

Water is lost by perspiration and sweating. Perspiration is the insensible loss of water vapour. It is continuous, uses the latent heat of vaporization and causes a loss of weight of about 0.5 kg in 24 hours. Vaporization is simple physical transpiration, not a function of the sweat glands.

The *sweat* is an acrid secretion, a weak solution of sodium chloride. The amount of water so lost is about 0.5 litres a day but can increase up to 12 litres in hot weather. Salt loss in a day from profuse sweating may be as much as 120 g, and this is important in *heat exhaustion*, where there is collapse with a normal or subnormal body temperature. It may be prevented by adding salt to the water drunk. This is quite different from *heatstroke* (see below).

The sweat glands are controlled by the nervous system, which regulates not the amount of water lost (this is the job of the kidneys) but the body temperature. Sweating depends on the difference between internal and external temperatures, the rate of heat production in metabolism, and the humidity of the air. The amount of humidity in the air may be a serious limiting factor if saturation makes evaporation of fluid water impossible. A minimal rise of body temperature by a third or half a degree sets the mechanism in motion. Strangely, the palms and soles of the feet are not so stimulated, though they are by the emotional state, so-called 'psychic sweating'.

TEMPERATURE REGULATION

The body is maintained at a constant temperature, though the outside level may vary enormously. It can be kept warm in cold surroundings because heat is produced in oxidative metabolism, and cool in hot surroundings by heat loss in sweating. A constant temperature implies an exact balance between loss and production. The former is a function of surface area and is relatively greater in children, whose surface is much larger than in adults in proportion to body weight.

The average body temperature taken in the rectum is 37.5°C, in the inner ear 37.5°C, in the mouth 36.5°C and in the armpit 35.5°C. But there is a daily fluctuation of one or two degrees, highest in the early evening and lowest in the small hours. This fluctuation has superimposed on it variations at different phases of the menstrual cycle in women. Organs like active muscles are hotter than the skin. There is a temperature gradient from the surface of the skin to the interior of several degrees.

The regulation of body temperature may be effected by changes in heat production or heat loss; both are simultaneous.

▶ Regulation of heat production

All the tissues, especially the muscles, produce heat. The liver is also a major heat generator as a result of its metabolic activity. With any change in environmental temperature, heat production is increased or decreased to compensate. In response to cold there is muscular activity, which may amount to shivering; at the same time liberation of adrenaline stimulates metabolism. External warmth reverses these changes.

These changes are due to the nervous system, which plays on the muscles, and by hormones such as adrenaline. Left to themselves, the tissues would behave like those of cold-blooded animals, producing *more* heat as the outside temperature rises; this is what actually happens when the regulatory mechanism fails, as in *heatstroke*.

▶ Regulation of heat loss

Heat may be lost by *conduction, convection* or *radiation*, or by *evaporation* from the skin surface. *Conduction* is of little importance as the skin and subcutaneous fat are insulators. *Convection* is more important. The air adjacent to the skin is warmed and rises, and each warmed layer is replaced by colder layers, which are warmed in their turn. *Radiation* is also significant since the body surface is a very efficient radiator.

Evaporation varies greatly with external circumstances. At low temperatures there is only insensible water loss; sweating begins at an air temperature of 29–30°C. At body temperature, sweating and evaporation are the only means of heat loss, since convection and radiation cease when there is no temperature gradient between skin and environment. The cooling effect of evaporation is due to the fact that energy is needed to convert the water of the sweat into water vapour, and this energy is abstracted from the body.

About two-thirds of the heat loss of a clothed adult is due to convection and conduction, less than a third to evaporation. These are physical considerations but heat loss is also controllable by physiological activities. Most important is fluctuation in the calibre of the small capillaries of the skin. These may open up to produce flushing, or *vasodilation*, which increases surface temperature and so accelerates heat loss, or they may shut down to produce a cold pallid skin surface – *vasoconstriction* – with reduced heat loss. The maximum blood supply to a patch of skin may be 50 times the minimum. These changes are under the remote control of the central nervous system, via the sympathetic and parasympathetic nerves.

This mechanism ceases to operate when the outside temperature reaches the level at which sweating and evaporation become significant. These are so efficient that a person can stand temperatures that will cook a piece of meat in a few minutes, provided the atmosphere be *dry*, so that evaporation is unhindered. If the air is saturated, temperature will rise rapidly, with possible death from heatstroke. Here there is damage to the vital centres of the brain, the regulatory mechanism breaks down, and the body temperature rises until death occurs.

Although sweating and skin blood supply are locally managed by the autonomic nervous system the ultimate control is from special centres in the brain, which are very sensitive to minute changes in the temperature of the blood and react to restore it to its previous level. They are very sensitive to the nervous stimuli that heat or cold receptors on the skin transmit.

The heat regulating centres work in the ways we have discussed: by producing shivering, sweating, more or less heat production and muscular activity, changes in metabolism, vasoconstriction or vasodilatation. The mechanism is usually set at a certain level and reacts to maintain that level under all circumstances. But if it is upset, the setting may change, and this is what happens in fevers, where it is adjusted upwards.

Clothing protects from extremes of temperature by providing a good insulating layer, with air entangled in its interstices. But clothing cuts down the radiating power of the body, and its permeation with moisture is responsible for the chilling effect of a cold damp climate.

HEAT STRESS

The first effect of heat stress is dilatation of the skin vessels. The skin becomes warmer and loses more heat. Second, there is sweating, which can stabilize the skin temperature at about 35°C. Very high temperatures can be tolerated if the air is absolutely dry. Tolerance lessens with increasing humidity, and air quite saturated with water vapour cannot be borne for long, even if the temperature is only 32°C.

Repeated exposure to high temperatures increases the ability to sweat, and with acclimatization comes a reduction in the salt content of the sweat. In excessive heat, sweat may be lost at the rate of over 4 litres an hour, though this cannot be maintained for long without collapse. In this *heat exhaustion* there is no fever and the skin temperature may be quite low.

Heatstroke is quite different and is collapse due to serious disturbance of thalamic function akin to the fever of infection. There may be severe hyperpyrexia and coma, possibly fatal.

EXPOSURE TO COLD

The immediate response to cold is by vasoconstriction of the skin vessels, which reduces the blood flow to the skin, whose temperature falls. At the same time heat production in muscle is stepped up by movement or shivering; the latter appears to be a more efficient source of heat than ordinary voluntary contraction. As far as longer-term acclimatization is concerned, humans tends to evade the problems rather than overcome them, by constructing a subtropical microclimate within special clothing and buildings.

Survival under conditions of exposure depends on the maintenance of body temperature, body water and supplies of energy, in that order of importance. Failure to maintain a proper temperature will kill before dehydration, and dehydration before starvation, though of course these factors interact.

If the body temperature falls and remains low, it is surprisingly easy to die of hypothermia in the middle of a civilized community, whether one is a fell walker, an older person living alone in an unheated room in winter, or a baby not kept warm. When clothing is soaked it loses its insulation value, and a person so clothed walking against a high cold wind might as well be naked, for he or she can only survive while he or she remains capable of generating a good deal of heat through physical work, e.g. walking or other exertion. Should the person begin to slow down, his or her surface temperature will fall, then the central organs begin to cool, consciousness will be lost at or even above 30–32°C, and death ensues in an hour or two unless these conditions are reversed by rapid rewarming of the whole body. Impermeable outer clothing is the great protection against cold, wet and the chill factor of wind velocity.

EXERTION: THE BIOCHEMISTRY OF MUSCLE CONTRACTION

The *immediate* source of the energy used in contraction is the splitting up of 'high-energy' adenosine triphosphate (ATP), which is involved in the actin-myosin reaction.

$$\text{ATP} + H_2O = \text{ADP (adenosine diphosphate)} + H_3PO_4 + 8 \text{ kcal}$$

This ATP has to be resynthesized, so energy is required from other sources. Under *anaerobic* conditions, when sufficient oxygen is not available locally, as at the beginning of high intensity exercise, substances in the muscle fibres break down to provide energy for this resynthesis. The most important is glycogen, its degradation product being lactic acid. If this accumulates in the muscle, as it may during static contraction, fatigue and inhibition develop. Normally, the excess lactic acid diffuses into the circulation and is removed by oxidation; some is also taken up by the liver for reconversion to glycogen. Some of the lactic acid remaining in the muscle is restored to glycogen during the recovery period after contraction, but much is burnt away to carbon dioxide and water by oxygen under *aerobic* conditions. The glycogen stores depleted during heavy prolonged exercise may not be fully restored for several days.

In a steady state of exercise under *aerobic* conditions, fat and carbohydrate brought to the muscles by the bloodstream provide energy for resynthesis of ATP. The liver is an important source of glucose, but even this may become depleted, and in prolonged exertion hypoglycaemia may develop (a concentration of glucose in the blood below normal) and fat becomes more important as a fuel. In prolonged exertion the muscle stores of glycogen repair the carbohydrate deficiency and the lactate level in the blood rises.

An adequate supply of oxygen is essential to contracting muscle, which may use ten times as much as in the resting state. The presence of lactic acid in the bloodstream during exertion stimulates the respiratory and cardiac centres, both to oxidize circulating lactic acid and to deliver oxygen to the contracting muscles. The oxygen is needed, not so much to enable muscle to contract as to allow it to recover from the effects of contraction.

The oxidative removal of lactic acid does not keep pace with accumulation at the beginning of continued effort, and its concentration in the blood rises. A main reason for breathlessness is to increase the oxygen supply, to get rid of excess lactic acid in the blood. When a steady state is achieved, the oxygen intake can deal with the acid as it is formed, and the individual gets a 'second wind' and can continue comfortably. However, after a short period of very strenuous effort, one is out of breath for several minutes, as there is an accumulation of lactic acid in the blood still to be oxidized. The body has got into 'oxygen debt' by accepting more calls on its oxygen supply than it can meet at the time.

The situation is complex because the carbohydrate store in muscle is important as a short-term source of energy during the relatively anaerobic conditions at the beginning of exercise, before the local and general circulation have had time to adapt, whereas the main source of energy during sustained activity is the free fatty acids of the blood. In a short burst, such as a hundred metres' sprint, most of the energy comes from the breakdown of muscle glycogen and a large oxygen debt is incurred, i.e. the circulation has not provided oxygen and nutrients locally just when they were needed. Increase of cardiac output, the redistribution of blood to the muscles and increased intake of oxygen by respiration are not immediate. Also, muscles used rhythmically, as in running, only have an intermittent blood supply because the contractions themselves empty the blood vessels. Again, this creates a partial oxygen deprivation and dependence on a diminishing local glycogen store. A step to overcoming this difficulty is the presence in muscle of a pigment, myoglobin, akin to the haemoglobin of the blood, which can liberate oxygen locally by dissociation. Oxygen uptake from respiration increases with the rate of exercise. The maximum uptake for young men is around 3.5 litres per minute, for young women 2–3 litres per minute, and for trained athletes 6 litres per minute. The oxygen content of the atmosphere, the capacity of the lungs and the work of the heart are all limiting factors. The muscles may use the oxygen supplied more or less efficiently, the arm muscles being less efficient than the leg muscles. Other factors are the efficiency of the circulation and the haemoglobin content of the blood. If less haemoglobin is available to transport oxygen, as in anaemia, or if the atmospheric oxygen is inadequate to saturate the haemoglobin, as at high altitudes, muscle efficiency will be impaired until there has been an adaptive increase in the number of red corpuscles.

Much *heat* is produced during muscular contraction, partly from the oxidative processes, partly from mechanical work. Only some 25% of our energy output can be translated into mechanical work; the rest produces heat. Most of the heat generated by exertion is dissipated by the evaporation of sweat. But body temperature rises during the first hour of exercise, and thereafter the raised temperature is maintained. Bodily mechanics are more efficient at higher temperatures; hence the importance of warming-up for athletes.

IMMERSION

The length of survival during immersion in cold water depends on the water temperature. If this is freezing, death occurs within the hour but may be delayed for up to six hours at 15°C. At 20°C or over, death does not occur from hypothermia. These facts may seem challenged by experience of swimming the English Channel, where individuals remain in cold water for much longer periods; but these subjects are exceptionally well insulated by fat and are trained to maintain a very high rate of energy production for many hours, and even they may succumb if they slow down. The ordinary shipwreck survivor does better to remain still because heat loss from movement exceeds any heat that can be produced by struggling.

The main problem of the survivor in an open boat is that of dehydration. It is not helpful to drink even small amounts of seawater, and death comes more quickly to those who do, because the concentrating powers of the kidney are limited to a urine containing the equivalent of no more than 2% of salt. Seawater contains 3–4% and drinking it is bound to raise the salt content of the tissue fluids with fatal results. The best course for survivors is not to drink at all for the first 24 hours, and then to ration any available fresh water at the rate required to compensate for obligatory water loss in urine, perspiration and expired air, i.e. 1 litre a day. The minimum intake should be of the order of 500 ml/day, as this maintains the general condition and ability to cooperate in rescue attempts. Water loss in hot weather can also be reduced by remaining still, shading the body and covering it with clothing soaked in seawater. But in a cold sea it is important to keep the clothing dry, as there is a risk of both hypothermia and immersion foot, a condition analogous to trench foot, marked by swelling, insensibility and muscle damage.

An individual – especially a child – who falls to the bottom of an ice-cold pond may survive if rescued after as long as an hour. This is because the glottis has closed automatically, so water does not enter the lungs, and the cold has suspended the brain's need for oxygen.

DESERT CONDITIONS

Here the problems are those of heat, dehydration and starvation in that order of importance. Heat may cause heatstroke and also increase water loss to the point of heat exhaustion. Water loss is much increased because the body not only ceases to lose heat through convection and radiation but may even gain heat from the environment. Under these conditions, only sweating affords any means of heat loss. Some rise of body temperature, even as high as 39°C, is acceptable. But at 40°C and over, impaired consciousness and collapse begin, while at 42°C, even if death does not ensue, there will be permanent brain damage. If it is possible it is best to crawl under something such as a vehicle and stay there.

Yet it is possible to survive almost any degree of heat if the victim keeps quite still and has reasonably adequate water supplies and, preferably, some shade. The effect of the sun's rays on the head and neck may be discounted. At night the desert may be intensely cold, and this is when some attempt to travel should be made. Death is usual when 15–20% of the body water, i.e. 10–15% of body weight, has been lost. The body of an average man holds some 40 litres of water and 2–4 litres may be lost without too much ill effect; after 5 litres mental and physical impairment set in and death occurs when 6–8 litres have been lost; such are the narrow margins of human existence on this planet.

HIGH ALTITUDES

In addition to the hazards of cold, dehydration (exacerbated by high winds) and starvation, high altitudes carry special risks arising from the reduced barometric pressure and consequent lowered availability of oxygen.

The oxygen content of both inspired air and arterial blood is reduced with circulatory and respiratory consequences. There is a stimulation of respiratory rate and volume and of cardiac output. A more delayed response, taking one or more weeks, is an increased production of red blood cells. Human populations habitually living at over 3000 m (10,000 feet) have a rate of red cell production 30% higher than normal and as much as 50% more haemoglobin than those living at sea level. Oxygen lack at heights may impair judgment and there is a risk of frostbite. Wind increases heat loss and more water is lost with the forced respiration, so clothing and shelter become important, even the insulating protection of a snowdrift.

The onset of altitude hazards depends to some extent on the rate of climb and the possibility of acclimatization. Mountaineers who ascend gradually to 4800 m (16,000 feet) may exhibit only slight breathlessness, and below 3000 m (10,000 feet) a fit young adult usually notices nothing untoward. Sudden exposure is much more serious, as in aircraft cabin depressurization or oxygen mask failure. The features of *acute mountain sickness* include headache, breathlessness, rapid pulse, weakness and sometimes vomiting, but usually pass off within a few days. *Chronic mountain sickness* is more serious and is due to reactive thickening and pressure rise in the arterioles of the lungs, with fluid accumulation there (*pulmonary oedema*) and in other tissues.

BAROTRAUMA

A diver is exposed to increasing pressure on all the tissues of the body at increasing depths. If he or she is breathing air, the partial pressures of nitrogen and oxygen driven into solution in the blood and tissues are increased, but not that of carbon dioxide because this is produced within the body. The increased nitrogen pressure may cause *nitrogen narcosis* at depths of 36 m (120 feet) and more, with unconsciousness at 105 m (350 feet).

On return to the surface the nitrogen may come out of solution as bubbles and cause *decompression sickness*. The gas blocks small blood vessels in the bones and nervous system and causes joint pains (the 'bends'), paralysis of the legs and difficulty in breathing. Large segments of the long bones may be infarcted and die, even years later. All these effects may also be seen in tunnel workers as 'caisson disease'. The treatment for these symptoms is immediate recompression in a pressure chamber, followed by a very slow return to atmospheric pressure. *Prevention* is by staged decompression at various depths on the way up, giving sufficient time at each level for gas to be eliminated without bubble formation. As this is tedious, divers are sometimes brought up in a submersible pressure chamber at high pressure, which is then gradually reduced on board ship. The dangers of deep diving can be lessened by using oxygen mixed with helium. It is unsafe to give pure oxygen because breathing this at high pressure risks the development of epileptic fits or severe bronchopneumonia. The use of pure oxygen in space capsules, even at low pressure, carries a serious fire risk. Pure oxygen given to premature infants has been linked to permanent damage to the eye.

It is interesting that oxygen given at increased pressure – about two atmospheres – has proved very useful in the medical treatment of certain conditions, notably gas gangrene, and is a useful adjunct to the X-ray treatment of cancer, the resuscitation of victims of coal gas poisoning, and the management of peripheral arterial injury or disease.

STARVATION

This is less serious than dehydration; survival is possible for as long as 60–70 days without any food whatever, provided that water supply is adequate and physical work kept to a minimum, though some work is good for morale. In starvation the body lives on its own tissues, mainly the fat reserve, but also to some extent on muscle protein. There is apathy and listlessness, giddiness and psychological withdrawal, also a fall in body temperature and basal metabolic rate. These changes are in part adaptive. The main factors in keeping alive are to maintain body temperature, water and calorie balance, in that order of importance. As little energy should be expended as possible. The value of even a small ration of food is out of all proportion to its caloric value. It is best not to include protein, so as to avoid overloading the kidneys and increasing the urine output if there is also a shortage of water. The most suitable ration is carbohydrate; alcohol can also be very useful.

Inherent biological rhythms

In contrast to the variations in the environment, the body has certain intrinsic rhythms in the way it functions. These are called *circadian* (about a day) as they are usually based on a 24-hour cycle. Thus body temperature is lowest in the morning and highest in the late afternoon, even in fevers. Urine output is lowest at night and urine is more alkaline in the morning. There are many biochemical variables, as in the level of plasma phosphate or cortisol. Therefore it is best to measure these at the same time of day if the readings are to be comparable.

These variations are independent of work or eating patterns up to a point, and evidently manifest the existence of an underlying 24-hour biological clock, unaffected by physical or mental activities or social pressures. The cycle can be made to adapt by altering circumstances, e.g. by working at night and resting by day, or flying into a different time zone, but adaptation may take weeks. Even in experimental subjects subjected unknowingly to artificially short or long days, many rhythms retain their 24-hour periodicity for weeks. However, in people exposed to darkness and solitude, as in caverns, the cycle gradually drifts into longer and longer periods of 25, 26 or 27 hours, until eventually several days may be 'lost'.

These rhythms are usually beneficial, e.g. wakefulness during working hours, sleepiness at night. The low urine output at night leaves sleep undisturbed. The higher temperature by day favours mental and physical work. It is interesting that the daily temperature variation often becomes inverted in night-workers, though with a reversion to pattern at weekends. The adrenal cortex secretes more cortisol in the early morning around waking, due to increased hypothalamic activity and pituitary output of ACTH, reflecting a change in brain activity and the return of alertness after sleep. It is fascinating that certain parasitic worms infesting the body also have their rhythms, as noted by their periodic migration between tissues and blood. Some favour the night to appear in the bloodstream, others the day; and these patterns are reversible if the host inverts his or her day/night pattern of living.

It is probable that most of the circadian rhythms are acquired during the first years of life under the influence of day and night alternation, and the pattern of human activities associated with this. One argument for this is that they are absent in Inuit populations, who are not exposed to such a regular alternation of light and darkness.

Melatonin is secreted by the pineal gland in the roof of the midbrain, with high levels at night and low levels in daylight. As the gland indirectly receives information from the visual pathways, these changing levels may be a means by which the day/night cycle affects the rhythmic physiological processes discussed above. Melatonin also seems to affect mating behaviour and gonadal size, inhibiting precocious puberty. It has been used to offset the syndrome of jet-lag.

Test yourself

1 What is the average volume of water lost each day through sweat?
   a 2 litres
   b 3 litres
   c 4 litres
   d 5 litres

2 Which method of heat loss occurs when the environmental temperature is the same as body temperature?
   a Conduction
   b Convection
   c Evaporation
   d Transpiration

3 What is the maximum uptake of oxygen for a trained athlete?
   a 3 litres per minute
   b 4 litres per minute
   c 5 litres per minute
   d 6 litres per minute

4 Which one of the following is only associated with *chronic* mountain sickness.
   a Headache
   b Breathlessness
   c Vomiting
   d Pulmonary oedema

5 Which gas can cause 'the bends' in divers?
   a Nitrogen
   b Oxygen
   c Carbon dioxide
   d Helium

6 How long can a person survive without food?
   a 30 days
   b 40 days
   c 50 days
   d 60 days

7 Which pigment in muscle helps to provide muscles with oxygen?
   a Haemoglobin
   b Myoglobin
   c Melatonin
   d Haemorrhage

8 What percentage of body water loss results in death?
   a 2%
   b 5%
   c 10%
   d 15%

9 Identify one adaptation of human populations living at high altitude that helps them to cope with the reduced partial pressure of oxygen.
   a   Increased breathing rate
   b   Increased heart rate
   c   Increased red blood cell production
   d   Increased sweat production

10 Identify one potential effect of breathing pure oxygen.
   a   Epileptic fit
   b   Headache
   c   Bronchitis
   d   High blood pressure

Test answers

Chapter 1	Chapter 3	Chapter 5
1 c	1 c	1 c
2 c	2 a	2 b
3 a	3 a	3 c
4 c	4 c	4 a
5 b	5 b	5 d
6 d	6 c	6 d
7 a	7 c	7 a
8 a	8 a	8 c
9 b	9 c	9 b
10 a	10 b	10 c

Chapter 2	Chapter 4	Chapter 6
1 c	1 b	1 a
2 b	2 c	2 b
3 a	3 c	3 d
4 b	4 a	4 b
5 a	5 d	5 d
6 a	6 b	6 b
7 d	7 b	7 b
8 b	8 a	8 b
9 a	9 d	9 b
10 d	10 c	10 a

Chapter 7
1 d
2 a
3 a
4 d
5 c
6 a
7 b
8 b
9 d
10 a

Chapter 8
1 d
2 b
3 d
4 c
5 d
6 a
7 c
8 a
9 a
10 c

Chapter 9
1 c
2 a
3 c
4 b
5 c
6 a
7 b
8 c
9 b
10 c

Chapter 10
1 c
2 b
3 c
4 b
5 d
6 d
7 d
8 b
9 a
10 a

Chapter 11
1 a
2 c
3 c
4 b
5 b

Chapter 12
1 c
2 b
3 c
4 a
5 d
6 c
7 c
8 b
9 a
10 c

Chapter 13
1 d
2 a
3 b
4 a
5 c
6 a
7 b
8 b
9 d
10 d

Chapter 14
1 c
2 b
3 a
4 a
5 c
6 c
7 c
8 b
9 a
10 d

Chapter 15
1 b
2 c
3 c
4 d
5 a
6 c
7 c
8 a
9 a
10 b

Chapter 16
1 c
2 d
3 d
4 a
5 a
6 b
7 a
8 d
9 b
10 a

Chapter 17
1 c
2 b
3 a
4 c
5 d
6 a
7 c
8 a
9 b
10 b

Chapter 18
1 b
2 a
3 b
4 d
5 c
6 b
7 c
8 d
9 a
10 d

Chapter 19
1 c
2 d
3 c
4 b
5 d
6 a
7 a
8 c
9 d
10 c

Chapter 20
1 b
2 b
3 c
4 c
5 d
6 a
7 c
8 b
9 a
10 d

Chapter 21
1 b
2 d
3 c
4 d
5 b
6 b
7 a
8 a
9 c
10 b

Chapter 22
1 c
2 d
3 c
4 a
5 c
6 a
7 b
8 b
9 a
10 b

Chapter 23
1 d
2 c
3 d
4 d
5 a
6 d
7 b
8 d
9 c
10 a

Index

Notes

Notes

Notes

Notes